Extra Dimensions in Space and Time

Multiversal Journeys

Itzhak Bars · John Terning

Farzad Nekoogar
Founding Editor

Extra Dimensions in Space and Time

Foreword by Lawrence M. Krauss

Springer

Itzhak Bars
Department of Physics & Astronomy
University of Southern California
College of Letters, Arts & Sciences
Los Angeles, CA 90089-0484
USA
bars@usc.edu

John Terning
Department of Physics
University of California, Davis
Davis, CA 95616
USA
terning@physics.ucdavis.edu

Founding Editor
Farzad Nekoogar
Multiversal Journeys
17328 Ventura Blvd.
Encino, CA 91316
USA
nekoogar@multiversaljourneys.org

Multiversal Journeys™ is a trademark of Farzad Nekoogar and Multiversal Journeys, a 501(c)(3) nonprofit organization

ISBN 978-1-4614-2532-8 e-ISBN 978-0-387-77638-5
DOI 10.1007/978-0-387-77638-5
Springer New York Dordrecht Heidelberg London

Softcover reprint of the hardcover 1st edition 2009

Springer is part of Springer Science+Business Media (www.springer.com)

To future scientists: Connor & Nima.

Farzad

To my family:

Itzhak

To: Laura, Jackson, and Maya.

John

Foreword

It is a pleasure to write this foreword for Multiversal Journeys' first book to further its goal of educating the public on the latest in theoretical physics results. I was privileged to participate in the lecture series on which this book is based, and enjoyed working, on separate occasions, along with John Terning and Itzhak Bars to carry on afternoon-long sessions with interested members of the public on issues ranging from extra dimensions to the future of the universe.

The lecture series was unique in that lengthy question and answer sessions were added to the lectures, to give the audience a chance to truly explore these fascinating areas in depth. The Multiversal book series is equally unique, providing book-length extensions of the lectures with enough additional depth for those who truly want to explore these fields, while also providing the kind of clarity that is appropriate for interested lay people to grasp the general principles involved.

As I have described in my own popular book, *Hiding in the Mirror*, and as both Itzhak Bars and John Terning colorfully describe in their reviews of the physics motivations for considering extra dimensions in this volume, humans have been fascinated for centuries by the possibility that there are extra dimensions, beyond the three spatial dimensions and one time dimension of our experience.

Over the past 90 years, beginning with the seminal work of Theodore Kaluza and Oskar Klein, this possibility has found its way into the heart of physics. The realization that gravity was associated with the curvature of space suggested that perhaps the other forces in nature might be associated with curvature in other, unobservable dimensions.

Starting in the 1960s an effort to deal with nasty infinities in quantum field theories pointed toward the possibility that fundamental elementary particles are in fact excitations of more fundamental units, string-like objects whose quantum vibrations would determine the spectrum of observed particles. It turned out that quantum strings, however, only make mathematical sense and only remove infinities if the number of space–time dimensions is much greater than four. In the original string models one required 26 dimensions!

Once supersymmetry was discovered as a possible symmetry of nature, it became possible to reduce the critical number of dimensions for viable strings to 10 or 11 dimensions. Since that time it has become clear that dimensionality itself may be illusory, and that what the physics encoded in a given number of dimensions may in fact be completely described by a projection onto one less dimension, just as a hologram encodes all the 3D information associated with an object on a 2D plate.

Of course, once the possibility of other space–time dimensions was proposed, the obvious question then became, where are they and how come we cannot detect them? Klein's first answer formed the basis of most of the subsequent work in this area. If extra dimensions of space are "curled up" on scales smaller than we can detect with our probes then they could effectively remain invisible.

More recently another more dramatic possibility was proposed. Gravity is by far the weakest force known in nature. If somehow all the other forces were restricted to our 4D manifold, and only gravity could propagate in extra dimensions then once again these dimensions could have been thus far undetected. Moreover, the size of the other dimensions would no longer be restricted to be small. They could be infinitely big!

In addition, the fact that gravity could propagate in more than three spatial dimensions also suggested a possible explanation for why gravity is so weak in our universe. Instead of a $1/r^2$ falloff that we see on scales in which we have thus far measured gravity, if gravity propagated in higher dimensions on smaller scales, it would experience a faster falloff. Thus, in fundamental scales the strength of gravity could be comparable to that of the other forces in nature and yet on scales we can measure it today it would appear far weaker.

One final exotic possibility remains. Highest dimensional theories involve more than three spatial dimensions but one time dimension. Some people have wondered whether it might be possible, in fact, to derive sensible laws of physics in a universe with higher dimensions and two or more "time-like" dimensions. This remained as an interesting but controversial possibility for a long time, but Itzhak Bars has solved these problems in his formulation of Two Time Physics with a new powerful gauge symmetry in phase space. He has been able to construct a physically meaningful framework that includes one extra time and one extra space dimension. As explained by Bars in this volume, successful physical theories in 3+1 dimensions now have counterparts in 4+2 dimensions. These theories make interesting predictions for observable physical phenomena that may extend our understanding of physics in three spatial dimensions and one time dimension.

Of course, it is necessary to add here that in spite of the remarkable theoretical interest in extra dimensions, and their great aesthetic appeal to many scientists and many members of the general public, no *direct* experimental evidence whatsoever yet exists for their presence (although Bars has argued that his Two Time Physics actually predicts observed relations in our world). Thus the question of whether they are remarkably interesting mathematical possibilities or whether they reflect an important underlying fundamental physical reality is currently not universally agreed upon by physicists. Of course, this is the price we pay when we explore the very limits of our understanding of nature.

Tempe, Arizona, USA
July 2009

Lawrence M. Krauss

Preface

There is a hidden sun in a particle
Suddenly the particle opens up its mouth

The earth and the heavens are cut into pieces
In presence of the sun escaping its trap

Rumi
1207–1273

One of the most popular topics in Multiversal Journey's lecture series has been the concept of extra dimensions in space and time. The topic has been covered in many of our conferences within the last few years: four times by the authors of this book.

The word dimension comes from the 14th century Latin *dimetiri* meaning to measure out. Historically, the notion of a dimension has long been used in geometry for centuries. The dimension of an object is specified by its coordinates (degrees of freedom: longitude, latitude, and height). In algebra, the notion of a dimension takes an abstract form such as dimensions of a vector space which are not the same dimensions we experience everyday in life.

In physics, the idea of extra spatial dimensions originates from Nordstöm's 5-dimensional vector theory in 1914, followed by Kaluza–Klein theory in 1921, in an effort to unify general relativity and electromagnetism in a 5-dimensional space–time (4 dimensions for space and 1 for time). The Kaluza–Klein theory didn't generate enough interest with physicist for the next five decades, due to its problems with inconsistencies. With the advent of supergravity theory (the theory that unifies general relativity and supersymmetry theories) in late 1970s and eventually, string theories (1980s), and M-theory (1990s), the dimensions of space–time increased to 11 (10-space and 1-time dimension).

In contrast, to the volume of research that has been conducted in the area of extra spatial dimensions in the last 40 years, not much time has been devoted to multi-dimensional time theories. Earlier attempts in multi-dimensional time theories had problems with violating causality among other issues. Two-time physics, a theory with 4-space and 2-time dimensions, is well covered in this book.

There are two main features in this book that differentiates it from other books written about extra dimensions:

The first feature is the coverage of extra dimensions in time (two-time physics), which has not been covered in earlier books about extra dimensions. All other books mainly cover extra spatial dimensions.

The second feature deals with the level of presentation. The material is presented in a non-technical language followed by additional sections (in the form of appendices or footnotes) that explain the basic equations and principles. This feature is very attractive to readers who want to find out more about the theories involved beyond the basic description for a layperson. The text is designed for scientifically literate non-specialists who want to know the latest discoveries in theoretical physics in a non-technical language. Readers with basic undergraduate background in modern physics and quantum mechanics can easily understand the technical sections.

The two parts of the book can be read independently. One can skip Part I and go directly to Part II which covers extra dimensions in space.

Part I starts with an overview of the standard model of particles and forces, notions of Einstein's special and general relativity, and the overall view of the universe from the Big Bang to the present epoch (Chapters 1, 2 and 3).

Chapter 4 covers basics of symmetry and perspective including local, global, and gauge symmetries.

Beyond the first four chapters, Part I covers two-time physics (Chapters 5, 6, 7, 8, 9, and 10). Two-time physics is the heart of Part I of this book and is best described by its author, Prof. Bars:

> Humans normally perceive physical reality in 3 space and 1 time dimensions and this is encoded in equations of physics in 3+1 dimensions (1T-physics). However, as discussed in this book, 1T-physics systematically misses to predict certain additional real phenomena in 3+1 dimensions in the form of hidden symmetries and hidden relations between apparently different dynamical systems.
>
> 2T-physics, which is based on a fundamental symmetry that can be realized only by adding an extra space and an extra time dimensions, is a completion of 1Tphysics that captures the missing information as effects due to the extra dimensions.
>
> According to 2T-physics in 4+2 dimensions, using 1T-physics in 3+1 dimensions is like analyzing only "shadows" on walls by observers stuck on "walls=3+1", while using 2T-physics is like analyzing directly the "substance" in the room by observers in the "room=4+2". The more powerful perspective of being in the "room" makes predictions, beyond those of 1T-physics, that can be tested and verified directly in our own 3+1 dimensional spacetime.
>
> With proper interpretation, some observations in our own spacetime can be used to peek into and explore the extra 1+1 dimensions which are neither small nor hidden.
>
> 2T-physics has worked correctly at all scales of physics, both macroscopic and microscopic, for which there is experimental data so far. In addition to revealing hidden information even in familiar "everyday" physics, it also makes testable predictions in lesser known physics regimes that could be analyzed at the energy scales of the Large Hadron Collider at CERN or in cosmological observations.

The technical sections in Part I are provided in the form of footnotes.

Part II of the book is focused on extra dimensions of space. As was mentioned earlier, one can skip Part I and go directly to Part II. It covers the following topics:

- The Popular View of Extra Dimensions
- Einstein and the Fourth Dimension
- Traditional Extra Dimensions
- Einstein's Gravity
- The Theory Formerly Known as String
- Warped Extra Dimensions
- How Do We Look for Extra Dimensions?

The technical section for Part II is covered at the end under, extra material: the equations behind the words.

I am indebted to Professor Lawrence M. Krauss for writing the foreword to the book. I would like to extend my thanks to the other two members of the advisory council for the book: Professor Mark Trodden and Professor David Finkelstein.

I would also like to thank the staff of Springer, especially Jeanine Burke and Dr. Harry Blom for making this project happen.

I am grateful to the following people associated with Multiversal Journeys who helped this project immensely:

- Multiversal Journeys, board of directors, professor Gene Moriarty, Richard D. Holt, and Dr. Faranak Nekoogar for careful review of the manuscript.
- Arthur Rieman, MJ's attorney for his legal advice and putting the book series contract together with Springer.
- Kai Barzilai for his technical support.

Finally, I would like to thank my father for providing the quote from Rumi and his encouragement for this project.

Multiversal Journeys — Farzad Nekoogar
Encino, California, USA
July 2009

Contents

Part I
Two-Time Physics: The Unified View from Higher Dimensional Space and Time

Itzhak Bars

Foreword to 2T-Physics

This part of the book aims at making the notions of 2T-physics more accessible to interested people. The exposition is mostly descriptive, but there will be some formulas (mostly in footnotes) to satisfy the curiosity of those who can follow them. The book is written at a pedagogical level that non-experts with little or no background in physics should be able to follow and enjoy for the most part. There are, however, some unfamiliar notions that may require more concentration when some of the more technical topics are discussed. Those topics will repeatedly come up, so it is likely that the ideas will sink in gradually as the reader makes more progress. Of course, readers with some prior knowledge on space–time, extra dimensions, string theory, or particle physics will be able to follow all aspects of the theory more easily. Physicists from every field as well as scientists or engineers with some mathematical background, who can understand the formulas, are very likely to get a considerable deeper insight on the ideas discussed in this book.

In this foreword I will first describe the notions of 2T-physics in a nutshell and then provide a road map to the various chapters.

Humans normally perceive physical reality as 4 dimensional, namely 3-dimensional space: up–down, back-forth, and side-to-side, plus 1-dimensional time: past–future. It is therefore natural that the ordinary formulation of physics, which I will call 1-time physics (1T-physics), has been developed in 3-space and 1-time dimensions. However, evidence has been accumulating that 3+1 dimensions are insufficient to describe our world, just like shadows on walls alone are insufficient to capture the true essence of an object in a 3-dimensional room.

Two-time physics (2T-physics) is compatible with 1T-physics, but it proposes a more fundamental formulation of physics in a 6-dimensional universe, comprised of 4-dimensional space and 2-dimensional time. Phenomena in 4+2 dimensions appear to observers in 3+1 dimensions the same way as an object in a room would appear as its own shadows on the surrounding walls. Naturally each distorted shadow, taken one at a time, contains considerably less data about its source. Similarly, 1T-physics in 3+1 dimensions describes each "shadow" one at a time and appears unable to capture all the information coming from 4+2 dimensions. Indeed there is more information in the real world in 3+1 dimensions, in the form of hidden symmetries and "dualities," that are missed systematically in 1T-physics. The hidden symmetries are distorted forms of higher dimensional symmetries and additional properties of higher dimensions inherent in each "shadow," while dualities are hard to notice relations among the "shadows." These additional phenomena are predicted by 2T-physics, while 1T-physics is only able to verify them but unable to predict them. The more encompassing and unifying 2T-physics is a completion of 1T-physics.

According to 2T-physics, using 1T-physics is like analyzing only "shadows on walls" by observers stuck on "walls," while using 2T-physics is like analyzing directly the "substance in the room" by observers in the "room." So, describing our common experiences by formulating them in the form of 1T-physics is like relying on restricted observers that are stuck in 3+1 dimensions and viewing a highly symmetric 4+2-dimensional system from a less favorable non-symmetric 3+1 perspective.

More generally, 2T-physics in d+2 dimensions is a completion of 1T-physics in (d–1) space and 1-time dimensions. The extra (d–1) – 3 space dimensions beyond the familiar 3 dimensions can be interpreted similar to the hidden dimensions of string theory. However, the additional extra space and extra time dimensions that takes us from 1T-physics to 2T-physics are neither small nor hidden; they are visible indirectly by proper interpretation of observations in our own 3+1-dimensional space–time.

While agreeing with the ordinary formulation of physics in 3+1 dimensions, 2T-physics reveals the hidden information as mentioned above. Some of those are easy to verify in macroscopic and microscopic physics as will be discussed in this book. In addition, 2T-physics also makes subtle predictions that make it distinguishable from other approaches in the description of lesser known regimes of physics. Therefore 2T-physics could also be tested through both computations and experiments at the level of particle physics at the energy scales of the Large Hadron Collider as well as in cosmology.

2T-physics emerges from requiring a new fundamental symmetry in the equations of motion in classical or quantum mechanics. This is a symmetry between covariant position (X^M = time & space) and contravariant momentum (P_M = energy & momentum). Such a symmetry is hinted in parts of 1T-physics, as in canonical transformations in the commutation rules that define quantum mechanics. Requiring that an appropriate extension of this symmetry must be obeyed at *every instant* in all equations that describe *all motion* is the most important new ingredient that uniquely leads to the equations of 2T-physics. The phenomena missed systematically by 1T-physics in 3+1 dimensions are captured by the requirement of this new *gauge symmetry* that acts on the phase space degrees of freedom (X^M,P_M) in 4+2 dimensions. The prediction of an extra time and an extra space dimensions as compared to those in 1T-physics is a direct consequence of this symmetry.

The 2T-physics approach provides new mathematical tools and new insights for understanding our universe at all scales of distance or energy. Evidence of the

4+2-dimensional world can be found at both the macroscopic and microscopic scales in the form of hidden symmetries and "dualities." Examples include the origin of symmetries that control the behavior of planets in celestial mechanics, the behavior of electrons in the hydrogen atom, and the behavior of quarks, leptons, and force particles in particle physics and cosmology. These are explained as hidden symmetries in 3+1 dimensions inherited from obvious symmetries in 4+2 dimensions, thus providing partial evidence for the existence of higher space–time.

Field theory is the most successful language so far for discussing the fundamental laws of nature, including the known four forces and the known fundamental matter in the form of quarks, leptons, and force particles. Applying the consequences of the new symmetry in particle mechanics to field theory has led to 2T-field theory, as a structure above 1T-field theory, with the power of making new predictions that are missed systematically in 1T-field theory. Amazingly, the best understood fundamental theories in physics that correctly describe nature as we know it today in 3+1 dimensions, namely the standard model of particles and forces and general relativity, have been shown to be "shadows" of more symmetric field theories in 4+2 dimensions. So, 2T-physics has worked correctly at all scales of physics for which there is experimental data so far.

The following is a road map to the various chapters.

To set in context the issues addressed and problems solved by 2T-physics, I have included some introductory material on past and current understanding of our universe from the very small to the very large. This includes a description of some of the historical evolution of notions on fundamental laws and principles, the current experimentally established structure of matter up to quarks, leptons, and force particles, the standard model of particles and forces, notions of Einstein's special and general relativity, and the overall view of the universe from the Big Bang to the present epoch. This material appears in Chapters 1, 2, and 3.

Since symmetry in physics and perspectives of observers are of fundamental importance to convey the notions of 2T-physics, I have included an elementary discussion of "global" symmetry and "gauge" symmetry in Chapter 4. It contains some discussion of how gauge symmetry appears in various theories including general relativity, the standard model, string theory, and 2T-physics. Gauge symmetry is the fundamental principle that leads to the existence and the detailed description of all the known forces and interactions. It is also the reason for the appearance of extra dimensions in string theory and in 2T-physics, as discussed in Chapters 5 and 6.

The shadows allegory is a helpful device to explain to the non-expert the relation between 2T-physics and 1T-physics. This allegory is discussed in Section 5.2 and is referred to in many later discussions in the rest of the book.

Although 2T-physics is already partially discussed in the earlier chapters, a systematic and considerably more detailed exposition is given in Chapter 7. This includes an informal explanation based on the shadows allegory and a formal discussion that displays some formulas. More details on the formulas and some related material appear in the footnotes. Consequences of 2T-physics, the missed information in 1T-physics, and the unifying aspects of 2T-physics are illustrated with one specific example that leads to Fig. 7.7. This figure captures concisely the main new notions of 2T-physics that apply not only to the specific example but also to all physical circumstances. In this way we see that 2T-physics is indeed a unifying structure that stands above 1T-physics.

In Chapter 8, some concrete examples of how to "see" 4+2 dimensions in everyday physics in 3+1 dimensions are discussed. Indeed, for the hydrogen atom and the celestial mechanics of planets, I show that observers in 3+1 dimensions can easily interpret their experimental observations as revealing information about the higher space–time in 4+2 dimensions. Other cases that appear in Fig. 7.7 are not as easy to describe with words or drawings, but with little mathematical effort and the guidance provided by 2T-physics, it is straightforward to learn about the presence of an extra time and an extra space dimension from observations directly in 3+1 dimensions plus the proper interpretation. These illustrate how we can *indirectly* see the effects of 4+2 dimensions.

Chapter 9 shows how to relate 2T-field theory to 1T-field theory through the shadow analogy, and in particular how to describe the standard model and general relativity in 2T-physics directly in 4+2 dimensions. The resulting shadow 1T field theories in 3+1 dimensions successfully describe all known aspects of fundamental physics as we know it today. But there are some new restrictions on scalar fields, such as the Higgs field and others that participate in cosmic phase transitions in the history of the universe. These new restrictions may help distinguish 2T-field theory from the usual less restrictive 1T-field theory through tests that may be conducted at the Large Hadron Collider at CERN or in cosmological observations.

Finally Chapter 10 covers recent progress as well as conceptual and philosophical directions to be pursued in 2T-physics. Progress in the past couple of years has extended from the 2T standard model and 2T-gravity to grand unification, supersymmetry, including higher dimensional super Yang–Mills theory up to 10+2 dimensions. The latter cannot be addressed by 1T-physics at all. Remaining challenges, the completion of the 2T-physics formalism for supergravity, strings, and branes, are also discussed. Future directions, which include the development of new computational techniques based on the hidden dualities and hidden symmetries, as well as conceptual and philosophical notions, are outlined.

For further reading, a complete list of my original papers on 2T-physics is included at the end of the book. The reader interested in more details should be able to find the desired information in those papers. Furthermore an index that points to the occurrence in the text of many physics or mathematics terms that may be unfamiliar to the non-expert should be helpful for both educational and informative purposes.

2T-physics sprouted in 1995 from my attempts to understand more deeply the hints for 2 time dimensions in the extended supersymmetry of M-theory. These efforts led in 1996 to my proposal of S-theory which was an algebraic approach to symmetries with two times unifying some duality aspects of M-theory. The attempt to find dynamics that overcomes fundamental problems with two times is what led to the new gauge symmetry. I would like to thank my past collaborator Kostas Kounnas who in 1997 at CERN participated in the initial stages of research for a gauge symmetry. That effort eventually led in 1998 to the gauge symmetry now used in 2T-physics. My former student Cemsinan Deliduman, postdocs Oleg Andreev and Djordge Minic, and colleague Soo-Jong Rey, who collaborated with me on a number of papers at the early stages of 2T-physics developing crucial notions, deserve special mention. More recent progress, including the all important 2T standard model and 2T-gravity, was sparked by numerous productive discussions with my recent students and collaborators Shih Hung Chen, Yueh Cheng Kuo, Bora Orcal, Moises Picón, and Guillaume Quélin, to whom I am deeply indebted.

I should also mention influential discussions on many occasions over the years with Edward Witten and John Schwarz who have encouraged me to pursue my 2T-physics ideas, and conversations with Joe Polchinski, David Gross, as well as my USC colleagues Nicholas Warner, Clifford Johnson, and Krzysztof Pilch who have challenged me with questions that led to better and firmer results.

Finally, many thanks to colleagues, friends, students, and family members who read the manuscript and made recommendations that improved this book. In this regard I would like to thank Professor Werner Däppen and Dr. Douglas Burke who provided the insights of physicists from another field, my students mentioned in the previous paragraph who corrected many mistakes, and especially my son James Barsimantov and my wife Annie Rosenschein-Bars who represented the perspective of non-experts and provided suggestions for significant improvements.

Los Anseles, California, USA

Itzhak Bars
June 14, 2009

Chapter 1
A First Look at the Known Universe

Among the foundational questions in physics and cosmology today is the deep mystery of why we live in a space–time of 3 space dimensions and 1 time dimension? Nowadays physicists ask seriously whether there are more space or time dimensions, and go further to suggest ways in which extra dimensions could play a role in determining the properties of the universe we observe today.

What leads us to even ask such questions and adopt such views?

For thousands of years people everywhere have asked questions that relate to space and time. This is a great journey of inquiry. *The Bible* begins with the story of creation that includes a description of how space and time came into being. Societies in many parts of the world provided a variety of religious, philosophical, as well as scientific explanations of the nature that surrounded them.

When Isaac Newton encoded the laws of mechanics in the 17th century in the language of mathematics, there seemed to be a clear vision of what space and time meant. But the mystery of space and especially time thickened in the beginning of the 20th century with Einstein's discovery of relativity.

In the 21st century M-theory advocates a unified theory of matter and forces based on 10- space and 1-time dimensions. Two-time physics envisions a greater unification in a space–time with one extra space and one extra time dimensions beyond those of M-theory. With its extra 1+1 dimensions, 2T-physics reveals hidden relationships among physical systems not captured by the ordinary formulation of physics in ordinary space–time at all scales of distance or of energy.

The universe around us is not as simple as it seems to be at first sight. Let's begin by imagining how a little boy, such as my grandson Isaac in Fig. 1.1, is likely to interpret his world if left on his own without guidance. Everyday he sees that it becomes light when the Sun rises and then it becomes dark when the Sun sets. The Sun, the Moon, and all the stars seem to be going around the Earth. Everybody would agree with Isaac on these observations that all of us can verify every single day with our own eyes. Armed with this assurance, soon he is likely to come to the conclusion, as the early man did, that the Earth is the center of the universe.

It seems so logical, but yet this is not true at all, as we know well today. What is missed at first sight is the rotation of the Earth around the south–north pole axis, which is sufficiently slow so as not to be felt by Earth's residents. For Earth's

I. Bars, J. Terning, Founding Editor: F. Nekoogar, *Extra Dimensions in Space and Time*, Multiversal Journeys, DOI 10.1007/978-0-387-77638-5_1,

Fig. 1.1 Isaac beginning to discover nature (Credits, Julie Culver.)

inhabitants, this self-rotation creates the illusion that the universe rotates around the Earth.

However, if Isaac got on a spaceship and landed on the Moon that also rotates around its own axis, the Moon would then seem to be the center of the universe! Of course, we can tell from our perspective that it is not so without any hesitation. In fact, Isaac would be able to see clearly from his spaceship that both the Earth and the Moon do indeed rotate around themselves. The early man's initial conclusion that seemed so solid is easily shattered by today's routine experiment of space travel. The observations of the Earth's residents were not wrong, but the interpretation was, so a more subtle theory of the universe had to be developed.

Nowadays in the space age we describe all of the above in a few lines, but historically it was extremely difficult to reach the correct conclusion. That the Earth is the center of the universe became the dominant view of all humanity for several thousands of years. This was reinforced by religious beliefs as well as by elaborate theories advocated by philosophers, astrologers, and astronomers. Observing the heavens and the stars was a serious undertaking in many societies, including the ancient Babylonian, Egyptian, Jewish, Greek, Arab, Aztec, and many European civilizations. These were done for both divination or religious guidance through astrology and sheer curiosity that eventually led to philosophical and scientific studies.

A mathematical theory of the heavens, with the Earth at the center of the universe, was developed by the Greek astronomer Ptolemy (AD 85–165) in the Roman province of Egypt. Astronomers at that time were puzzled by the motion of some bright objects in the night sky which they called "planets," meaning *wandering stars*. Planets were distinguished because they moved across the sky in relation to the other stars. The motion of some planets, which at times seemed to change their smooth trajectories and begin to move backward for a while and then again forward, was baffling. In the *Almagest*, Ptolemy developed a complex mathematical scheme based on geometry that accounted for much of the astronomical movements of the seven planets known at that time. In this theory, a large rapidly rotating outer sphere contained the stars, while planets were embedded in smaller rotating spheres, one for each planet. The theory was especially convincing because a precise mathematical description of planetary motion around a static earth apparently fitted well the observations. Ptolemy's theory of the universe, represented in the model of Fig. 1.2 with the Earth at the center, became the dominant view in the Western world for 13 long centuries.

Fig. 1.2 Ptolemy's universe (Credits, INMA.)

However, additional planets were found eventually and those were harder to describe with Ptolemy's spheres. The scheme had to be revised and became too elaborate and less and less convincing. Finally the Polish astronomer Nicolaus Copernicus (1473–1543) proposed the *heliocentric* theory, that the Sun, and not the Earth, was the center of the universe, as seen in Fig. 1.3. The Earth was not static but, like the other planets, it moved uniformly around the Sun. This revolutionary approach could account more simply for the strange planetary motion as observed from the Earth, because the planets did not all move at the same rate. Copernicus described the planetary orbits as a superposition of circles.

Galileo (1564–1642), an Italian physicist, mathematician, and astronomer, supported Copernicus' view. He was one of the originators of the scientific method, which required quantitative experiments whose results could be analyzed with mathematical precision. He believed that the laws of nature must have a mathematical description. He made important contributions to the understanding of the laws of motion under the influence of gravity and other effective forces. He also built a small telescope and with it he discovered that the planet Jupiter had four moons circling it. This shattered Ptolemy's universe, since by this verifiable observation the Earth could not be the center for Jupiter's moons. It was no longer a matter of just theory, the correct model was experimentally evident! However, many astronomers and philosophers who believed firmly in Ptolemy's model, initially refused to believe

Fig. 1.3 Copernicus' universe. (Credits, NASA.)

that Galileo could have discovered such a thing. Galileo was accused of being a heretic by the Catholic Church, because the church, influenced by the literal interpretation of the Scriptures, also held the view that the Earth had to be the center of the universe. Galileo was put under house arrest by the Inquisition. To be able to save his life he had to recant, and he died in disgrace, condemned by the church.

The Copernican Revolution created a new perspective which led eventually to the German astronomer and mathematician Kepler (1571–1630) to develop the laws of planetary motion. Kepler had a much simpler mathematical description of each planetary orbit as being a well-defined single ellipse with the Sun sitting at one of its foci. Kepler's laws provided one of the foundations for Isaac Newton's theory of *universal gravitation*. By universal gravity, Newton meant that the gravitational force experienced on the Earth (like a falling apple) has the same origin as the gravitational force that holds the planets around the Sun. Newton was able to derive mathematically the elliptical planetary orbits from his laws of mechanics that followed from the simple statement

$$Force = Mass \times Acceleration,$$

where the force was due to the gravitational attraction of a relatively small planet to a large central Sun.

Except for small corrections that were explained later by Einstein's theory of general relativity, Newton's laws basically explain the motion of all objects in the solar system.

There are a lot more phenomena in the universe beyond the solar system, how are they described? By the end of the 19th century, physicists and astronomers had become convinced that it was possible to understand everything in the universe by using only Newton's laws of mechanics, his theory of universal gravity, and Maxwell's laws of electricity and magnetism.

However this order began to change right at the beginning of the 20th century with the dual discoveries by Planck and Einstein. In 1902 Planck proposed the quantum of energy which later led to quantum mechanics as the correct way to describe motion for subatomic distance scales. In 1905 Einstein proposed special relativity for the correct description of fast moving objects and in 1916 he introduced general relativity for the correct description of the gravitational force. General relativity includes extreme situations, when the gravitational force becomes very strong, such as the inside of black holes, or the time of the Big Bang which marks the birth of our universe.

Guided by these new insights, our horizons of the size and scope of the universe changed dramatically during the 20th century. This came about through further experimental discoveries made with large telescopes that explored remote parts of space far beyond our own galaxy, and with powerful accelerators that looked deep into the structure of matter in the subatomic and subnuclear regimes.

Hubble's observations of galaxies in the 1920s made it evident that the solar system is only a tiny part of the Milky Way galaxy. A galaxy is a conglomeration of billions of stars of various sizes similar to the Sun, which are held together as a dynamical system by gravitational attraction. Today it is estimated that there exists at least hundreds of billions of, maybe several trillion, galaxies in the universe.

It was at first thought that the matter contained in each galaxy is the same basic matter we see on the Earth, namely matter made up of quarks and leptons and the stuff that holds them together as explained in the next section. Not only that! According to later discoveries which refined this picture, there is more than meets the eye. As shown in Fig. 1.4, within each galaxy there is some additional matter that does not shine, which is called dark matter. The familiar matter has been estimated to make up only about 3–5% of the total energy of the universe, while dark matter accounts for 22–25% of it. The remaining almost 70–75% of the total energy is called dark energy, which is different than both ordinary and dark matter.

Even though we are aware of their existence, the nature of both dark matter and dark energy is poorly understood today. Dark matter responds to the gravitational force like ordinary matter does, but it does not seem to respond to the other familiar strong, weak, and electromagnetic forces. A hypothesis called WIMP, which stands for weakly interacting massive particles, attempts to partially describe dark matter. On the other hand, dark energy does not even behave like matter particles at all, but it still responds to the gravitational force although in a somewhat different way.

Starting with Hubble in the 1920s, it has been established that the universe as a whole is in motion and in a state of expansion. This is not understood in terms of Newton's laws. It is Einstein's general relativity that comes into play to explain it. Everything we observe today was all created in a Big Bang which initiated the expansion of the universe.

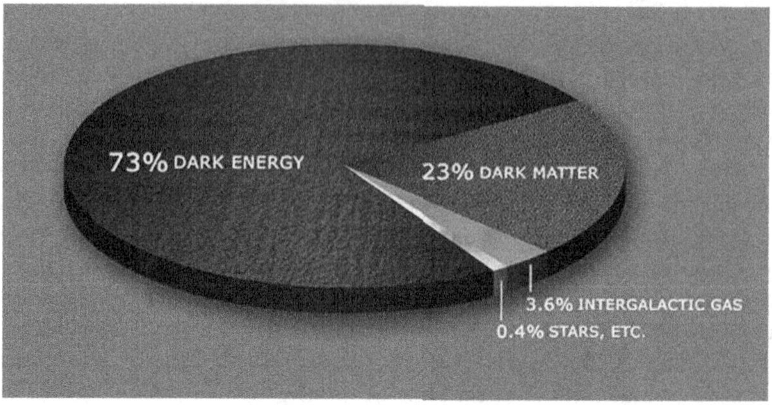

Fig. 1.4 Ratios of dark matter, dark energy, and ordinary matter (Credits, Particle Data Group.)

In recent years it has been established that the continuing expansion is not happening at a uniform rate. The expansion is actually slightly accelerating, that is, its rate is increasing due to some kind of outward pressure or energy. The origin of this effective energy is not currently understood, so it has been named "dark energy".

A possible theoretical explanation for dark energy is a modification of general relativity that Einstein called the "cosmological constant." For many decades we used to think, according to measurements, that the cosmological constant was zero. However, the observed acceleration in the expansion of the universe seems to indicate that it is not zero but is small and of the order of 10^{-29} g cm^{-3}. This implies that a tiny constant energy density fills every centimeter-cube of the entire universe and that there is an associated outward pressure of that magnitude that tends to expand each bit of volume. This energy–pressure density is very small and negligible for most phenomena that occur in a *finite amount of space*, but yet it is large enough to account for almost 75% of the total energy of the *entire* universe and to influence its rate of expansion.

Understanding the smallness of the cosmological constant has been a challenge for theoretical physicists for many decades, well before the recent experimental discoveries. If one tries to use naively the standard well-understood fundamental theories, that work well at the subatomic and large scales, to also estimate the size of the cosmological constant, one comes up naturally with a huge number which is wrong by a factor of 10^{120}. It is believed that this is an important message that something fundamentally important is missing in our understanding of the underlying theory of the universe.

In summary, today we believe in an expanding, slightly accelerating, universe that has no center! We, that is, humans and all living species, are far from being special or being at the center of this big and still mysterious universe.

Chapter 2
Structure of Matter and Fundamental Forces

The previous section provided a rough description of the large structures in the observable universe. We now go inside matter.

2.1 The Fundamentals at Subatomic Distances

When we look inside matter, at sizes much smaller than our natural sizes, as we go deeper and deeper into the subatomic scales, we discover entire new worlds of molecules, atoms, nuclei, quarks, and leptons. This is illustrated in Figs. 2.1 and 2.2.

During its early stages right after the Big Bang, the entire universe is imagined to be extremely dense, extremely energetic, and extremely hot. According to our current understanding of cosmology, a tiny speck of the early universe will eventually develop to become our present *visible* universe. So, to understand the status of our visible universe at the time of the Big Bang, you need to imagine squeezing all the mass and energy of the present universe into an unimaginably small region. During such early stages of the universe, elementary matter collides with huge energies within volumes 10^{20} times smaller than the sizes of nuclei or protons depicted in Figs. 2.1 and 2.2. Therefore the evolution of the universe from the Big Bang into what it is today must have been determined by the fundamental laws of physics that govern the smallest elementary particles, namely quarks, leptons, and force particles, moving in extremely small regions at huge energies. This is well beyond the levels of energies in so-called high-energy physics experiments at modern accelerators. So, we need to look deep into the structure of matter and understand thoroughly its elementary constituents and the fundamental forces acting on them, in order to explain our origins.

It has been experimentally established that all known matter in our environment is made up of quarks, leptons, and "force particles" that hold them together, as illustrated in Figs. 2.1, 2.3, 2.4 and 2.6. The behavior of these elementary constituents is controlled by four fundamental forces. The strong and weak forces are short range and can be felt only by tiny particles moving at subnuclear distances. However, the electromagnetic and gravitational forces are long range, and act at both small and large distances, including macroscopic distances typical of our everyday life.

I. Bars, J. Terning, Founding Editor: F. Nekoogar, *Extra Dimensions in Space and Time*, Multiversal Journeys, DOI 10.1007/978-0-387-77638-5_2,

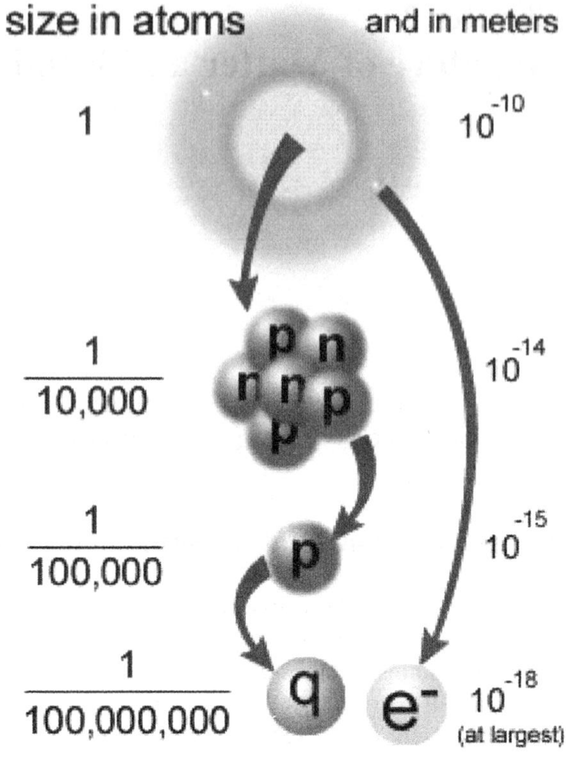

Fig 2.1 Structure of matter (Credits, Particle Data Group.)

For this reason we are more easily aware of the electromagnetic and gravitational forces.

The strong force, whose influence dominates within matter of the size of a proton (10^{-15} m in Fig. 2.1), holds quarks and gluons inside protons and neutrons and other similar strongly interacting particles called hadrons. The strong force is transmitted by the gluons.

The weak force has an even smaller range of influence (within 10^{-17} m) and it is responsible for the decay of matter, such as *neutron → proton+electron + anti-neutrino*. It is transmitted by the $W^{\pm}$ and Z^0 particles.

The electromagnetic force is responsible for holding the electrons around a nucleus in atoms (10^{-10} m in Fig. 2.1) and also for the formation of molecules from atoms. The range of the electromagnetic force is infinite; therefore, we can experience it also in the macroscopic world, such as when it acts to pull magnets together, or the attraction/repulsion of electric charges, as well as in all phenomena that involve light. The electromagnetic force is transmitted by photons, which are the smallest bits of light. Photons make up the electromagnetic waves in the entire spectrum of frequencies, including radar, radio, TV, X-rays, and the visible spectrum that is interpreted by the human eye as the colors from red to violet light.

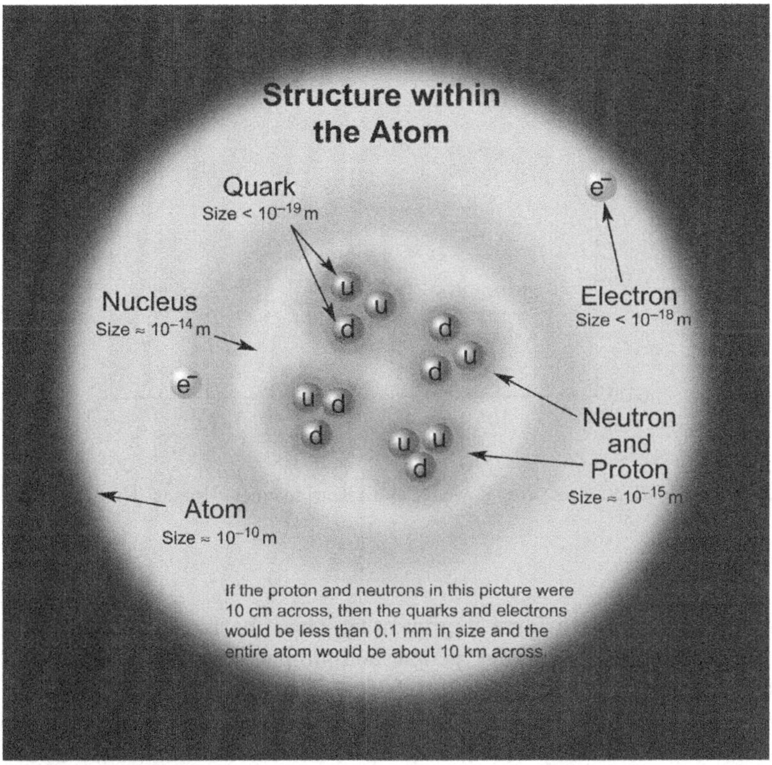

Fig. 2.2 Atom, nucleus, proton, neutron, *u*, *d* quarks, and electron (Credits, Particle Data Group.)

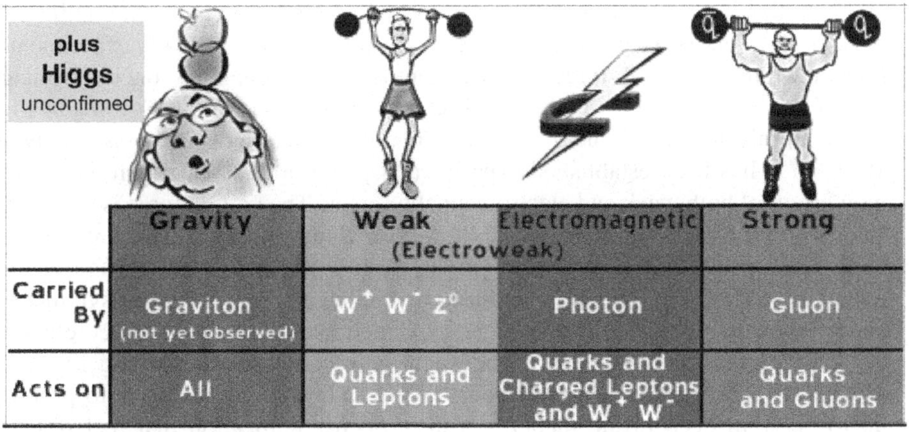

Fig. 2.3 The four interactions, force particles, and Higgs (Credits, Particle Data Group.)

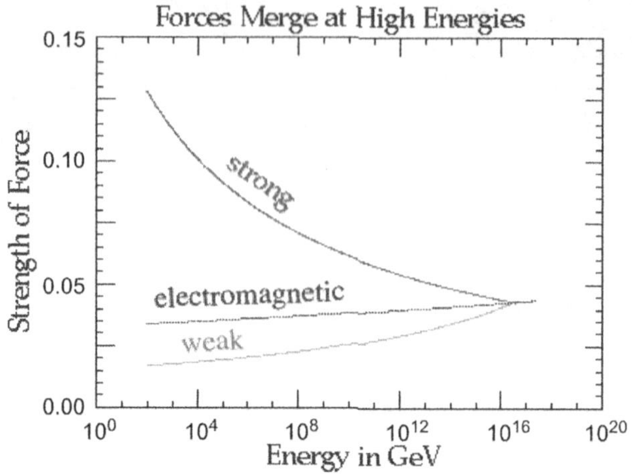

Fig. 2.4 Grand unification of strong, weak, and electromagnetic forces (Credits, Particle Data Group.)

The remaining known force is gravity which is the most familiar in everyday life. It is responsible for holding people and animals on the surface of the Earth, or holding planets around the Sun or other stars, or controlling the motion of stars within galaxies, or of the motion of galaxies within the universe. In fact, the gravitational force controls the expansion of the entire universe starting from the very initial Big Bang. The graviton is the particle responsible for transmitting the gravitational force.

Properties of the known forces are summarized in Figs. 2.3 and 2.4. Although described as separate forces with different roles in today's universe, there is some evidence that these forces possibly arise from a common origin and are unified at some deep level. As we go deeper inside matter through higher energy experiments, it is known that the weak force becomes comparable in strength to the electromagnetic force when particles collide with energy about 100 GeV, as shown in Fig. 2.4. With this amount of energy we can probe the structure of matter as deeply as 10^{-18} m. It has been established experimentally that at that scale there is a common origin of both weak and electromagnetic forces. The unified force is called the "electroweak force" and is described by a precise mathematical structure called a Yang–Mills gauge field theory based on the gauge symmetry SU(2) $\times$ U(1), as formulated by Weinberg, Salam, and Glashow. The meaning of a "gauge symmetry" and symbols like SU(2) $\times$ U(1) will be explained later in this book. Theoretically, a "grand unification" is envisaged between the electroweak and strong forces at around 10^{17} GeV in energy as shown in Fig. 2.4, which corresponds to probing up to 10^{-32} m deep inside matter. An even deeper unification that includes the gravitational force is expected at around 10^{-35} m, known as the Planck length, a typical length scale at the Big Bang, which is probed with energy of about 10^{19} GeV.

In addition to the force particles associated with the familiar four forces, there is an additional hypothetical particle called the Higgs particle as shown in Fig. 2.3. The Higgs particle is so far mysterious. It is needed to understand the origin of mass for all elementary particles; however, its precise nature remains to be clarified experimentally, which could possibly occur during the years 2009–2011 in experiments to be conducted at the Large Hadron Collider at CERN . The Higgs could be simply a single additional elementary particle or it could be analogous to the tip of an iceberg (Fig. 2.5) waiting to reveal a lot more of the secrets of the universe, including the possibility of new forces acting at deeper levels inside matter. We will return to the Higgs particle in a later section.

Turning now to the matter particles, quarks, and leptons, shown in Fig. 2.6, can be compared to the bricks that make up a house. By contrast, the force particles described above can be compared to the mortar that holds the bricks together. Both the bricks and the mortar come in several varieties, thus making up the properties of the various parts of the house as well as giving its overall character.

One of the distinguishing properties of elementary particles is their spin, which is a property similar to the amount of rotation of the Earth around its own axis. All quarks and leptons, and their anti-particles which make up anti-matter, have spin 1/2. On the other hand, the force particles have integer spins: the graviton has spin 2; the gluons, photon, $W^{\pm}$, Z^0 all have spin 1; the Higgs particle has spin 0. All half-integer spin particles are called *fermions* in honor of the great physicist Fermi. Similarly, all integer spin particles are called *bosons* in honor of another great physicist, Bose. Fermions and bosons behave in certain distinguishing ways according to the laws of quantum mechanics as well as according to the roles they play in the structure of

I got you a bunch of stuff. This is just the tip of the iceberg

Fig 2.5 Higgs particle, the tip of an iceberg? (Credits, Brian Zaikowski.)

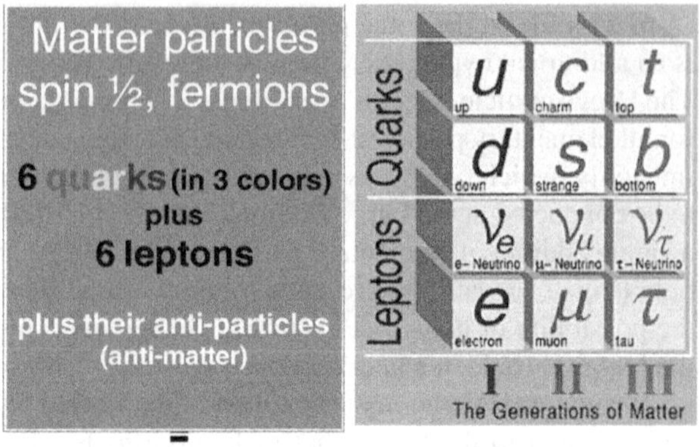

Fig 2.6 Three generations of quarks and leptons (Credits, Particle Data Group.)

matter, and this provides a basis for classifying them in separate sets of elementary constituents.

These "elementary" particles have various quantum numbers, which generalize the concept of the electric charge, that whimsically are called "flavors" and "colors." The various quarks and leptons are distinguished from one another by their colors and flavors, somewhat like distinguishing different ice creams in a Baskin–Robbins store, or like distinguishing people or animals from one another by certain characteristics.

There are six flavors of quarks called up (u), down (d), strange (s), charm (c), bottom (b), and top (t). These silly sounding names developed historically in the process of building models to characterize in words certain physical and mathematical properties that summarized experimental observations. There are also six flavors of leptons called electron (e), muon (μ), tau (τ), and the corresponding neutrinos called electron-neutrino ν_e, muon-neutrino ν_μ, and tau-neutrino ν_τ. There are three color charges: red, blue, and yellow. All quarks come in every variety of color. Therefore there are altogether 18 distinct quarks that can be distinguished by their flavor and color charges ($6 \times 3 = 18$). On the other hand the six leptons have no color charges, they are color neutral.

Anti-matter is as fundamental as matter. For every quark, lepton, or force particle, there exists also a corresponding anti-particle that carries the opposite values of the flavor and color charges (anti-up, anti-down, etc.). This is a prediction made by Paul Dirac on the basis of quantum field theory, before anyone knew of the concept of anti-matter. Today it is commonly observed in accelerators that, at sufficiently high-energy collisions among matter particles, both matter and anti-matter particles are produced, just with the predicted properties. Why don't we see much anti-matter in our common experience? Where is your anti-self-made of anti-matter? There are plausible, but not fully settled, cosmological explanations for why in today's universe the amount of anti-matter is greatly suppressed compared to the amount of matter, as further explained in later sections.

The 6×3 quarks and the 6 leptons make certain patterns of flavor and color charges that are repeated 3 times. This repetition defines the concept of three generations of quarks and leptons as organized in Fig. 2.6. The members in the third generation are heavier than the corresponding members in the second generation and those are heavier than the corresponding ones in the first generation.

The concept of generation is also related to how all of these particles interact in certain symmetric patterns with each other under the influence of the strong, weak, and electromagnetic interactions. The symmetry patterns have a mathematical structure[1] called SU(3) × SU(2) × U(1), where 3 stands for triplets of color SU(3), 2 stands for doublets of flavor SU(2), and U(1) is another structure related to both electromagnetic and weak interactions. The doublet structure is made evident in Fig. 2.6, where the quarks and the leptons of each generation come separately as doublets of flavor, and triplets or singlets of color, such as

$$\underbrace{\begin{pmatrix} u \\ d \end{pmatrix}_L^{red}, \begin{pmatrix} u \\ d \end{pmatrix}_L^{blue}, \begin{pmatrix} u \\ d \end{pmatrix}_L^{yellow},}_{\text{triplets of color}} \quad \underbrace{\begin{pmatrix} \nu_e \\ e \end{pmatrix}_L}_{\text{singlets of color}} \quad \left.\right\} \begin{matrix}\text{doublets of flavor, } L \text{ stands} \\ \text{for left-handed chirality}\end{matrix} .$$

The color triplet structure of SU(3) dictates precisely how quarks interact with the eight gluons that create the strong force. Similarly the doublet/singlet structure of SU(2) × U(1) dictates precisely how all quarks and leptons interact with the $W^{\pm}$, Z^0 and the *photon* that create the weak and electromagnetic interactions, respectively. Only the left-handed chirality components of quarks and leptons interact with $W^{\pm}$, while Z^0 and the photon interact with both the left and right chirality components. "Chirality" is related to spin, such that, for a *massless* particle it coincides with "helicity." Helicity is that component of spin in the same direction (right handed) or in the opposite direction (left handed) of its momentum.

Why there are only three generations, and why we see these peculiar patterns of flavors and colors within each generation, remains as an unsolved mystery. There are some tentative proposals to explain these patterns as a property of higher dimensions in string theory.

The mass pattern of the observed quarks, leptons, and force particles is rather uneven and cover a fairly big range from almost zero for neutrinos, to 5×10^{-3} GeV/c^2 for the electron, to 175 GeV/c^2 for the top quark. The exactly zero masses of the graviton, photon, and gluons are explained by gauge symmetry (discussed later), but the non-zero mass patterns of quark and leptons so far has defied an explanation.

Closely related to the *origin of mass*, an additional particle called the Higgs particle, or some more complicated structure that imitates it, is postulated to exist. According to theory, the patterns of interactions of the Higgs particle with all of

[1] In the Cartan classification of symmetry groups outlined in footnote 1 of Chapter 5, the group SU(3) corresponds to A_2 and the group SU(2) corresponds to A_1, both taken in their compact versions. The group U(1), which corresponds to just phase transformations, is not included in the Cartan's list of "simple" Lie groups. All of the "simple" groups are non-Abelian, while U(1) is Abelian, thus explaining why it is not included.

the other particles parallels precisely the observed mass patterns. There are indirect effects of the Higgs particle that explain several observations, and in this sense there is indirect evidence for its existence. But the Higgs particle has not yet been seen directly in accelerators because it is presumably too heavy to be produced with the available accelerator energies.

However, the new Large Hadron Collider (LHC) at CERN, that will begin to conduct crucial experiments in late 2009 or early 2010, will have sufficient energy to produce the Higgs particle. We don't really know whether what we call "Higgs" is just a single isolated particle or whether understanding it will uncover deeper structures. Seeing the Higgs, and clarifying its nature, is expected to be among the very first triumphs of the LHC.

Currently we don't know the true sizes of the quarks, leptons, and force particles force particles (graviton, photon, W, Z, gluons), but we do know that their sizes must be smaller than 10^{-18} m. This is 1000 times smaller than the size of the proton or neutron as indicated in Fig. 2.1. The LHC will be able to probe into distances 1000 times smaller than that and learn more about the properties of these tiny structures.

Are these particles point-like without any structures inside them or do they contain some smaller and more elementary constituents? There are several theories about that.

The most popular point of view is that quarks, leptons, and force particles are all made up of tiny strings. We will discuss some aspects of string theory later. String theory tries to answer simultaneously other puzzles of the universe, so it can serve as a useful guide when we don't have definitive answers. However, even if string theory is completely right, the strings may be so tiny that even the LHC would not be a sufficiently powerful "microscope" to detect them.

There are string-related "brane-world" scenarios that are more hopeful about the prospects of seeing string theory structures at the LHC. This will be discussed in the second part of this book authored by Prof. Terning.

The fundamentals of string theory are still under development so its attractive proposals are not yet firm conclusions. There are also less popular but nevertheless viable possibilities of inner structure, such as "preons" bound by a new strong force, that the LHC could discover.

Therefore we are still uncertain what, if anything, is inside quarks, leptons, or force particles. Perhaps the LHC will shed new light on this very important issue, but maybe not?

The computational framework of all known interactions in Figs. 2.3 and 2.4, including the Higgs, are precisely given by the *standard model of particles and forces*. This is a theory that has a mathematical structure dictated by *relativistic quantum Yang–Mills field theory* in 3-space plus 1-time dimensions. Each word here is loaded with deep mathematical and physical meanings.

"Relativistic" refers to the fact that it is a theory in the framework of Einstein's theory of special theory of relativity in 3-space and 1-time dimensions. Some aspects of relativity will be discussed in a later section.

"Yang–Mills field theory" is the local gauge symmetry framework that extends Maxwell's theory of electromagnetism to include and unify the other known interactions. The fundamental role of local gauge symmetry will be explained later, and this will be essential in the construction of two-time physics in 4 + 2 dimensions.

"Quantum" implies that all aspects of quantum mechanics, that accounts for the correct rules of motion with some probabilistic features at subatomic distances, have been incorporated, as long as gravity is ignored.

Quantum effects of gravity are indeed ignorable up to the energy level of all experiments conducted so far. It is only at much higher energies, far beyond those accessible in accelerators, that quantum effects of gravity can play a significant role to explain the physical phenomena (see later Fig. 9.1 and footnote 1 in Chapter 9). Within the present "low-energy" limitation of accelerators or other available detection capabilities, Einstein's theory of general relativity at the classical (rather than quantum) level has agreed perfectly well with all aspects of gravity measured so far.

It is known that at the theoretical level general relativity poses some severe problems in the context of quantum mechanics. Even though this is not of experimental consequence at the energy level of current or foreseeable accelerators, at the much larger Big Bang energy level it is a stumbling block that prevents us from deciphering the very origin of the universe. For this very essential reason, as well as a matter of principle, quantum gravity is a major theoretical problem that requires solution. String theory appears to be the only viable framework that could deliver the answer according to current understanding.

The precise formalism of the standard model, as a relativistic quantum field theory, provides the tools to perform computations and make quantitative predictions that extend our understanding beyond just a qualitative description of what is observed. It is a beautiful, compact, and simple structure that makes thousands of detailed predictions all of which are in exquisite agreement with very precise measurements so far. Some of these computations and measurements are so accurate that they agree up to 12 decimal places. There has not been a single experiment that has contradicted the predictions of the standard model up to now. The standard model of particles and forces is an amazing feat of the 20th century.

Despite its enormous success, the standard model leaves many questions unanswered. We will discuss these later.

2.2 Large Distances and Cosmology

We know that the laws of physics that govern the subatomic world seem to be rather different from the laws of physics that govern the big universe. Quantum mechanics and special relativity are part of the rules that govern the universe at the small scale. Actually the correct rules that apply everywhere, large or small, as we know them today, are the laws of particle physics in the subatomic world. It can be shown that the laws that apply at larger scales arise from the fundamental ones as an approximation that is effectively valid for the large structures. So, the *fundamental* laws of physics are unique and the same everywhere; they should not be confused with the effective approximations.

The fundamental laws are incorporated as part of the standard model of particles and forces outlined above. This correctly describes the subatomic world with great precision and great success. The standard model is experimentally verified and is today the most precise theory ever known to mankind.

Since the universe was so tiny during its early stages starting with the Big Bang, it must have been governed by the same laws of physics at small distances, as prescribed by the standard model. By applying this well-established knowledge we can give a very detailed description of what happened at each stage of structure formation described in Fig. 2.7. This makes a quantitative prediction of what we should expect to see in today's universe. The great success is that the prediction matches the experimental observation in quantitative detail as outlined below.

The details of the Big Bang itself still needs clarification, but the physics and history of the universe, starting right after the Big Bang, is pretty much under control.

During its early stages, right after the Big Bang, the entire known universe was very dense, so the elementary constituents of matter were squeezed into distances much smaller than the sizes shown in Figs. 2.1 and 2.2. They were moving at huge energies, much larger than those attainable in today's accelerators. There was so much energy that matter and anti-matter were continually created from energy and also under collisions they continually destroyed each other back into energy (all according to Einstein's formula, $E \Leftrightarrow mc^2$). The forces acting on matter were not sufficient to overcome the very energetic motion; therefore the quarks, leptons, force particles, and their anti-particles were roaming around almost as free (that is, not bound) particles. As shown in Fig. 2.7, as the universe expanded and cooled down, the existing matter, moving at lesser energies, started to coalesce under the influence of the strong, electromagnetic, and finally gravitational forces.

By applying the established laws of physics up to the level of quarks and leptons, and with some additional educated theoretical guesses, physicists can trace the evolution of the universe quantitatively starting as soon as 10^{-45} s after the Big Bang, as outlined below partly following the events depicted in Fig. 2.7.

Some thinkers like to identify the Big Bang as the beginning of time, at least for our universe, but so far we really do not understand the meaning of space–time at the Big Bang. The universe was then extremely small, dense, hot, and energetic. The laws of physics at those extreme conditions are not fully understood yet; therefore, some of the current views about what exactly happened could change as we learn more in the future. It is thought that matter and anti-matter were created in equal amounts from the energy available in the huge explosion ($E = mc^2$).

Shortly after the Big Bang, asymmetry developed in favor of matter due to asymmetric interactions known to exist among elementary particles as measured in the laboratory. This is why in today's universe we see mostly matter as illustrated in Fig. 2.8. There are remaining questions. The precise mechanism for how the matter–anti-matter asymmetry originated is still under debate. In the cosmological evolution, did matter/anti-matter asymmetry start with the quarks (this is called baryogenesis) or with the leptons (this is called leptogenesis) or both?

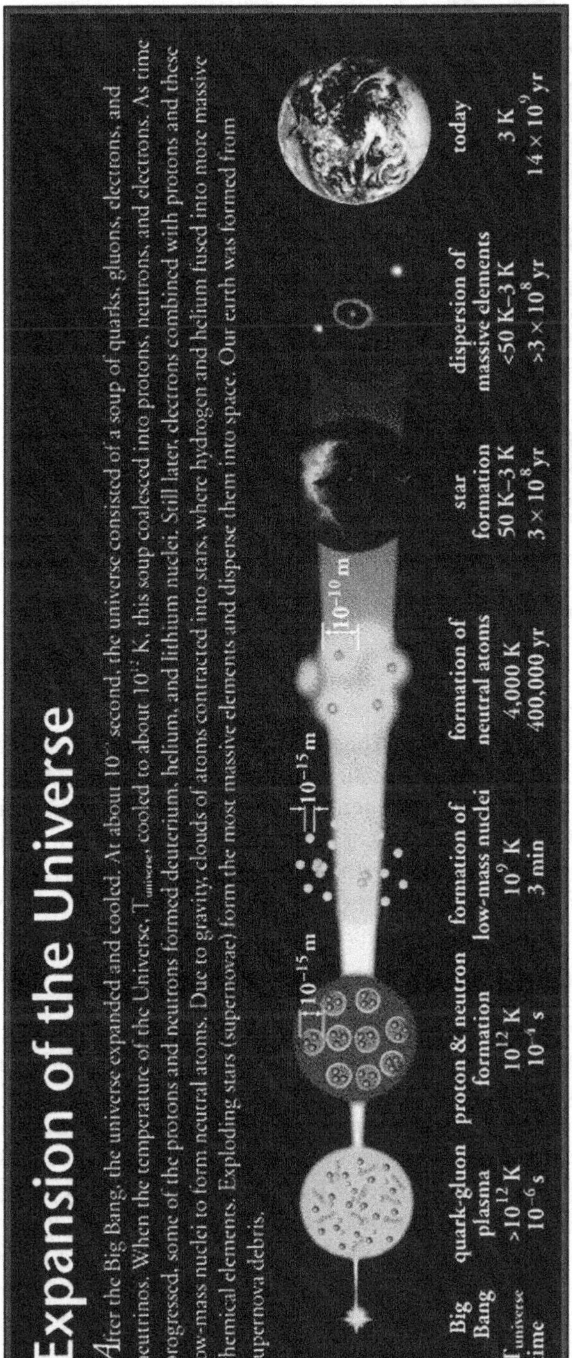

Fig. 2.7 Expansion of the Universe (Credits, Particle Data Group.)

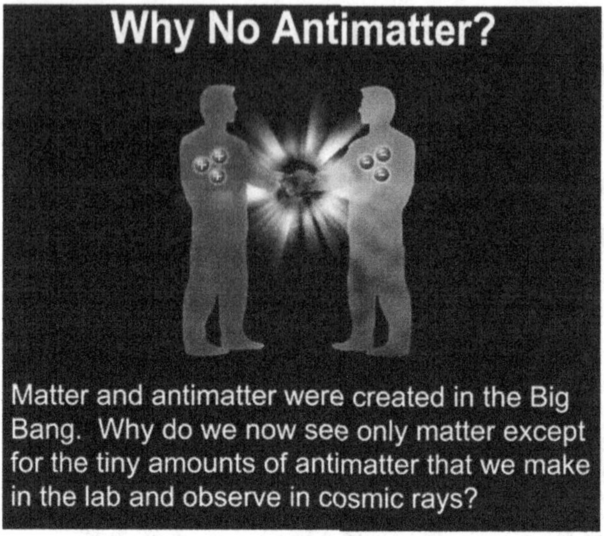

Fig. 2.8 Anti-matter is produced in accelerators when matter particles collide with each other at very high energies. But anti-matter is not found in abundance in the universe like matter is (Credits, Particle Data Group.)

Also a phase transition is thought to occur that split the unified force into three components as shown in Fig. 2.4. These eventually became the strong, weak, and electromagnetic forces that act on matter as recognized in today's universe.

The size of the universe underwent a sudden exponential inflation, like a bubble, at *speeds much larger than the speed of light*, which is permitted under the influence of strong gravitational fields as described by general relativity. This happened perhaps around 10^{-35} s after the Big Bang. Today we live in the space of what used to be only a speck of the original universe just before inflation. The fact that our "bubble" developed from a speck explains why the energy distribution, or temperature, in today's universe is isotropic and homogeneous and looks the same everywhere we look in the sky. It is because within a speck there is not much possibility to have significant variations in the energy distributions. Conceivably, there are other universes disconnected from ours in other inflation bubbles that may have developed from other specs with possibly different conditions than ours. Such other bubble universes are disconnected from ours because we cannot observe them. Our ability to observe is limited by the speed of light. In today's universe, the gravitational fields are non-existent in what is mostly empty space, so no signal can travel faster than light. Hence, we are unable to obtain information from other parts of the original universe that moved away from our bubble at speeds larger than light's. This inflation scenario is formulated mathematically in the context of general relativity.

After inflation, matter was in the most elementary form we see today in accelerators, including the quarks, leptons, gluons, photons, which we discussed above. These elementary particles were still so energetic that they managed to roam around almost freely, since the attraction produced by forces could not be overcome due to the great speeds.

Around 10^{-6} s after the Big Bang, while the universe was as hot as 10^{12} degrees, it was like a soup of somewhat interacting quarks and gluons, while photons, electrons, neutrinos, and other weakly interacting particles continued to move freely.

At around 10^{-4} s, under the influence of the strong force generated by the color charges of gluons and quarks, color neutral bound states that are called *hadrons* were formed. According to the theory of quantum chromodynamics those color neutral hadrons contain (quark + quark + quark) or (anti-quark + anti-quark + anti-quark) or (quark + anti-quark), where each quark or anti-quark can be any of the six flavors (u, d, s, c, b, t) in Fig. 2.6. Indeed these are the only types of hadrons that have been observed in accelerators. Most of these hadrons are short lived and decay quickly after they get created in collisions, but the proton which contains (up + up + down) and the neutron which contains (up + down + down), and their anti-matter counterparts are survivors. A free neutron lives about 1000 s while a free proton has never been seen to decay. Their sizes are about 10^{-15} m.

At around 3 min, or 10^9 degrees in temperature, the strong force pulled together the protons and neutrons to form small stable nuclei 10^{-14} m in size. Inside the nucleus, the neutron is slightly less massive and then it can no longer decay. So these small nuclei survive basically forever.

After about 380,000 years, at much cooler temperatures and slower motions, the electromagnetic attraction created by the positively charged nuclei managed to finally capture the, by then more sluggish, negatively charged electrons to form neutral atoms of 10^{-10} m in size.

Photons cannot bounce off neutral matter. Therefore once the matter coalesced into neutral atoms, the existing photons continued to move freely and filled the expanding universe. These relic photons are the origin of the *cosmic background radiation* that is detected today everywhere throughout the sky in the form of a background temperature of 2.7 K distributed *uniformly* across the universe.

The degree to which this temperature is homogeneously and isotropically distributed everywhere puzzled cosmologists who proposed that at a much earlier stage the universe underwent the rapid expansion mentioned above. *Inflation* explains the uniformity of this temperature distribution simply by making it plausible that we come from a small speck that could not have much variation within that small distance.

Modern telescopes can measure tiny fluctuations in this background temperature, as shown in Fig. 2.9. This image is captured by the satellite telescope called Wilkinson Microwave Anisotropy Probe (WMAP). It represents the temperature distribution across the sky in all directions. Colors have been exaggerated to represent very tiny deviations from the average 2.7 K, with blue indicating cooler temperatures. These precision measurements can be related to certain additional details in cosmological observations that confirm the theory of inflation.

The electromagnetic force cannot act on *neutral* matter over distances larger than molecules. So, after the formation of neutral matter, gravitational attraction eventually dominated over the other forces in shaping the universe. This led to the

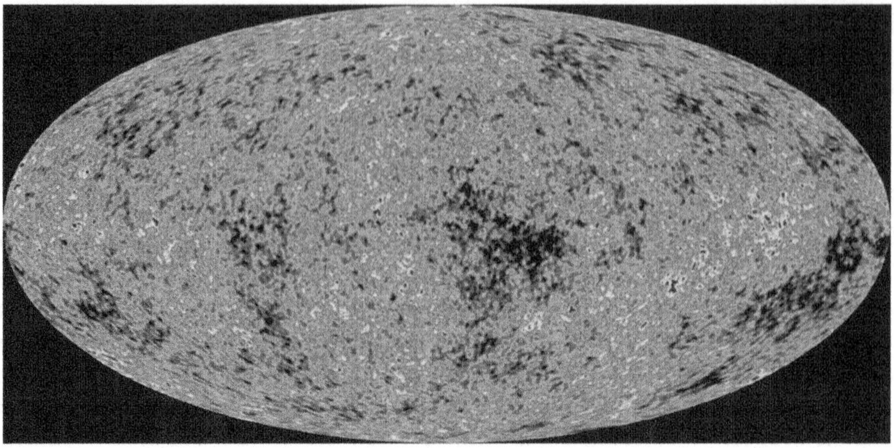

Fig. 2.9 WMAP image of temperature fluctuations (Credits, Particle Data Group.)

formation of the first stars and galaxies by starting from slight inhomogeneities in the distribution of energy. According to the theory of star formation, to reach today's state of the universe, the required amounts and distribution of inhomogeneity are in agreement with the tiny temperature fluctuations measured by WMAP in Fig. 2.9, and as predicted by the theory of inflation. Such inhomogeneities slowly coalesced in parts of the universe under the influence of gravity, leading to the condensation of hydrogen and helium into stars, and then into galaxies, after about 1–300 million years.

In the core of stars, gravity can create sufficiently huge pressures to fuse small nuclei into larger ones through processes called *stellar nucleosynthesis*. Nuclear fusion reactions at the cores of massive stars create more massive nuclei in a series of nuclear reactions. The heavier elements, up to as heavy as iron (Fe, with 26 protons and 30 neutrons), are formed in this way inside stars. The formation of the heavier elements requires more violent processes produced in star-like structures called red giants or astronomical processes called supernovae to fuse the smaller nuclei into larger ones. This is why iron and nickel are among the more abundant heavy elements found in the cores of planets such as the Earth and in metallic meteorites.

Stellar nucleosynthesis is the origin of the energy of the Sun and other stars. As long as there is enough matter to burn into energy ($E = mc^2$), the outward pressure created by the nuclear reactions keeps the remaining matter from collapsing under the influence of gravity.

When an aging star can no longer generate sufficient energy from nuclear fusion, it may undergo a gravitational collapse because the outward pressure from the nuclear explosions diminish. Through this collapse, and related gravitational processes, *red giants*, *white dwarfs*, or *neutron stars* are created. Spinning neutron stars may also form *pulsars*. If the star is too large, its life ends with a *supernova* explosion before becoming a neutron star and then finally a *black hole*. As measured today, in a galaxy the size of the Milky Way, supernovae occur about once every 50 years.

There may be black holes of various sizes roaming around a galaxy, but the small ones are hard to detect. However, large black holes create a large gravitational attraction that affects the motion and the distribution of stars in a galaxy. In fact, every galaxy is expected to contain at least one large giant black hole at its center. Experiments conducted during 1995–2003 established that the giant black hole at the center of our own Milky Way galaxy has an estimated mass of 4 million times the mass of our Sun and measures about 23 million km across, which is smaller than the orbit of Mercury around the Sun. Evidently a huge amount of mass is packed into the small region that makes up this black hole. In 2004 astronomers detected another intermediate-sized black hole of 1300 solar masses near the center of the Milky Way.

In the history of the universe, the expanding shock waves from supernovae explosions scattered the heavier elements into space where they later combined with interstellar gas to produce new stars and their planets.

Over time, clusters of galaxies formed and then superclusters formed, etc., leading to the larger structures in the universe, including strings of galaxies, sheets of galaxies, and large voids where there are nearly no stars.

Today the diameter of the observable universe is estimated to be 28 billion parsecs (about 93 billion light-years). This diameter is increasing at a rate of about 1.96 million km/s, which is about 6.5 times faster than the speed of light in empty space.[2] The age of this huge universe is estimated as 13.73 billion years since the Big Bang, with an uncertainty of about 120 million years.

The solar system, the Earth, and all its occupants, animate and inanimate, are the products of nuclear astrophysical processes followed by biological ones. The Sun formed about 4.6 billion years ago. The oldest rocks found on the Earth give an estimate of 4.54 billion years for the age of the Earth. Life developed in an elementary form in a primordial earth under special circumstances some 3.5 billions years ago. The first organisms made up of many cells appeared about 1.8 billion years ago, while it took mammals until about 200 million years ago to develop. Modern humans emerged much later, some 30,000 years ago.

---

[2]This expansion rate is calculated from the Hubble parameter $H_0 = (\dot{R}/R)$ whose value today is measured as $H_0 \simeq 70.1$ (km/s)/Mpc, where 1 Mpc $= 3.08 \times 10^{19}$ km. The inverse of the Hubble constant gives a rough estimate of the age of the universe. One finds $(\mathrm{H}_0)^{-1} = 0.44 \times 10^{18}$ s, or equivalently 13.9 billion years. Inserting the diameter of the universe today, $2R = 28{,}000$ Mpc, one obtains from the formula $2\dot{R} \simeq 1.96 \times 10^6$ km/s. This faster than light motion is possible in curved space–time, or equivalently under the influence of the gravitational force.

Chapter 3
What Is Space–Time?

Time has been discussed in philosophy, metaphysics, and physics. It continues to be a confusing notion to many people, especially because of a lack of understanding of Einstein's relativity theories, which are often misinterpreted and confused with the subjectivity of observers or even their psychological states. It has even been questioned by some whether "time" exists, that it may be only an illusion created by our minds.

Perplexing questions linger, such as does time flow in only one direction? Does it have an origin? Some such questions are clouded by our incomplete understanding of the Big Bang and what space–time may or may not be just at, or even before, the Big Bang. In the absence of a complete theory that answers many other related issues on quantum gravity, there is some room to question the fundamental reality of space–time at that level. This remains in the realm of speculation, and it would be premature to jump to conclusions.

However, the well-established aspects of physics give a clear answer about the nature of space–time starting soon after the Big Bang. Within that realm let us discuss what physicists mean by space and time.

There seems to be nothing confusing about time when referred to in connection to a series of events. For example, you wake up in the morning at 7:30 AM and take a taxi at 8:00 AM to go to the airport to catch a flight at 11:15 AM. If time does not exist what are these numbers that you measure with your watch? If watches are not synchronized with all others, how will everyone know that the plane will take off at exactly 11:15 AM? There seems to be a *universal time* in our everyday life that plays a major role in organizing sequences of events in a precise way that everyone agrees on. This sequence can also be recorded in a movie strip and run both forward and backward. Each point of this sequence is specified with a value of time. This everyday experience of a universal time is the notion of time advocated by Newton and is the one shattered by Einstein. We will discuss Einstein's view later.

In all aspects of established physics, including the standard model and general relativity in 3 + 1 dimensions, time is as simple and as real as space, although not as simple as it appears in everyday life. It is measured with standardized instruments just like space is. Both time and space are required to describe any motion, and they

I. Bars, J. Terning, Founding Editor: F. Nekoogar, *Extra Dimensions in Space and Time*, Multiversal Journeys, DOI 10.1007/978-0-387-77638-5_3,

both play their essential roles in the mathematical formulas that encapsulate the laws of nature.

It has been discovered that the fundamental laws of nature have certain symmetries, some of which we will discuss later. Among those symmetries there are some space–time symmetries that put space and time on an equal footing. However, this does not prevent different observers from having different perspectives of space and time that can influence what they see.

We discussed above that it was different perspectives that led to the apparently conflicting theories of Ptolemy versus Copernicus. But after 13 centuries we finally understood how to reconcile those perspectives.

Einstein showed that different *time* perspectives that deviate from the notion of universal time can also be created when observers are in relative motion or when they are under the influence of various forces. Under such circumstances they would measure different amounts of time for the same sequence of events even by using standardized instruments and no matter how objective they try to be. The symmetries teach us how to relate observers with different space–time perspectives to one another and how to insure that they correctly interpret their different observations while all are being described by the same set of laws of physics.

To clarify these concepts, and later expand them in new directions, let us first discuss the space–time perspective of a given observer, at rest.

In everyday experience the notions of space and time are intuitively obvious. As an example, consider an event that happens somewhere, at some time; for example catching a bus at the corner near your house, today 1-h after waking up. *Where* the event happened is related to the notion of *space* and *when* it happened is related to the notion of *time*.

To be clear about what we mean by "where" and "when" for an event, we will define an *origin* of space–time and, relative to that origin, we will define space and time *intervals* that specify the event. Consider again the event of catching a bus described above. For this event, the origin for space is your house and the interval in space is the straight line segment between your house and the bus stop. For the origin of time we will use the moment you woke up, and then the interval in time is the 1 h that passed until you caught the bus. To tell your friend *where* you caught the bus, you first refer to your house and then the space interval between your house and the bus stop. Similarly, to tell the time, you first specify when you woke up, and then the time interval up to the moment of the event.

The notion of 3-dimensional space is easy to grasp since we live and travel in it. As above, we can define the *origin* of space as the location of the observer and set-up an orthogonal system of coordinates at that location, as in Fig. 3.1. This observer can now report unambiguously the location of any point in 3-dimensional space by providing three numbers (x, y, z) that correspond to longitude, latitude, and height, which are called the space *coordinates*. A specific value of the coordinates is called the *position* of something, such as a particle or some event denoted by the ⋆ in Fig. 3.1. The *position represents the space intervals* in 3 dimensions relative to the origin. All possible positions where you can place the ⋆ in Fig. 3.1 is the

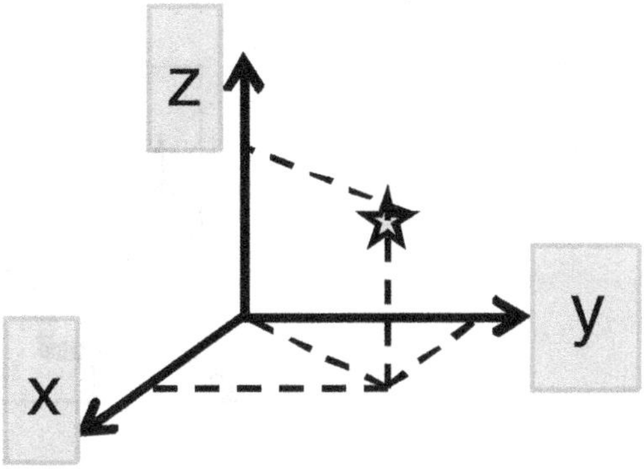

Fig. 3.1 Position in 3 dimensions

entire 3-dimensional space. So all of space amounts to all possible values, positive or negative, of the coordinates x, y, z.

In a similar way we can define some convenient origin of time for our observer, relative to which he will report time intervals, t, for various events. For future events t is positive, while for past events t is negative.

To specify an event, our observer must give us four numbers (t, x, y, z) to be able to say precisely when and where the event occurred relative to the origin of his space–time. The four intervals (t, x, y, z) must be measured by our observer by using standardized clocks and rulers so that everyone can interpret precisely when and where the event occurred.

Time appears as the *fourth coordinate* of space–time since four numbers (t, x, y, z) are required to specify any event relative to the origin. Pictorially we can represent 4-dimensional space–time, with a simplistic picture as in Fig. 3.2, where an event is represented with a ⋆ and its space–time coordinates relative to the observer at the origin are indicated. All the points where the ⋆ can be placed is the entire 4-dimensional space–time.

Next consider a series of events and plot them as a collection of space–time points that make up a curve as shown in Fig. 3.3. This curve is the *space–time history* that encapsulates both the past and the future, just like a movie strip does. The infinite set

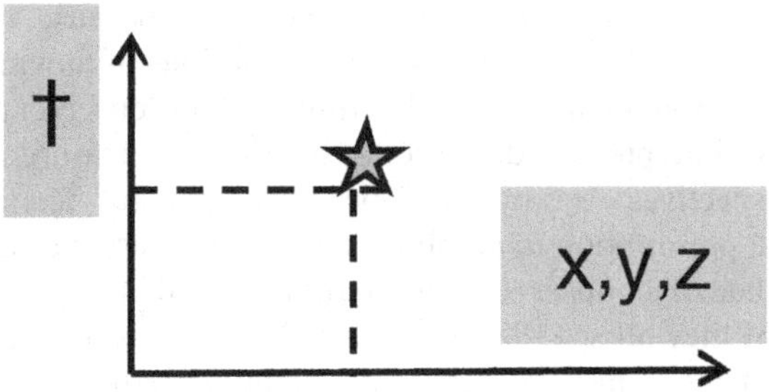

Fig. 3.2 An event in space–time

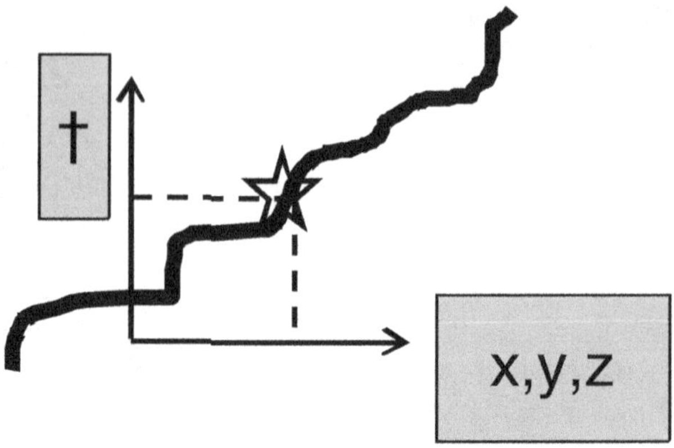

Fig. 3.3 History is a collection of events

of space–time points that make up a continuous curve representing the space–time history of a moving point-like particle is called a *worldline*.

The description of space–time and the worldline in Fig. 3.3 is relative to one *stationary* observer located at the origin. How do other observers see the same moving particle?

Imagine another observer whose origin is somewhere else in space–time. Then, the worldline in Fig. 3.3 will appear displaced relative to the new origin, but will maintain its shape. If in addition, the second observer keeps changing his space–time location, then the curve that represents the worldline will also be altered. The alteration in the shape of the curve can be smooth or erratic depending on the movements of the second observer. In particular, the time intervals between specific events as measured by the second observer can be different in comparison to the first one. Under certain circumstances of his motion, his observations can be so dissimilar that even the sequence of events may not be the same; that is, he could disagree about which event came first and which came next. So, observers that are in relative *motion* could report the same events with quite different measurements of both space and time *intervals* relative to their own *origin*. Hence, neither space nor time measurements are absolute; they depend on the observer. This is not because of any biological or psychological state or other subjectivity of the observers. It happens due to relative motion despite the fact that all observers use standardized equipment objectively.

So, there is no universal space–time because, relative to a static observer, space–time is redefined by the various motions of moving observers.

How then can we determine which observer to believe? How will we be able to interpret the disagreeing space–time measurements obtained from different perspectives even by the same trusted person? The answer is that we need to provide a prescription to all observers that measure the same set of events for how to translate other observers' measurements to their own perspective, and only then compare if they all see the same thing. If after this translation, all observers, in all possible states of motion, agree on certain properties, then those properties are universal to all observers, and obviously we can trust them.

The "dictionary" for the translation to various perspectives is provided by the symmetry properties of those laws. Finding the fundamental symmetries is perhaps the most fascinating part of theoretical physics.

It is in this symmetry sense that the *laws* of physics in any space–time are universal, although space–time itself is not universal, because the form of the equations for fundamental laws does not change as space–time changes, as shown by Einstein in the case of general relativity.

3.1 Einstein's Relativistic View of Space–Time

The discussion above applied to all possible motions of observers, including erratic or accelerated moves under the influence of various forces.

To describe the theory of special relativity, Einstein concentrated on what he called *inertial* observers, namely only those that move uniformly at *constant velocities* relative to each other, without any acceleration or change of direction. How does their state of motion influence their measurements of space–time *intervals* between events? What is the correct dictionary between such observers, so they would agree on the interpretation of their disparate observations.

Through his very simple but ingenious reasoning, Einstein was able to determine that Newton's laws of motion could not be compatible in principle with certain observations. Experiments proved him right, and then a new era of relativistic laws of motion began.

The essential input in Einstein's reasoning was his hypothesis that the speed of light, $c = 300{,}000\,\text{km/s}$, is universal for all *inertial* observers, as predicted by Maxwell's theory of electromagnetism. This seems to contradict common experience in everyday life, which is why it was hard to discover. In everyday experience, if you walk on the street at the rate of 2 miles per hour, and then walk at the same rate on a long train which itself is moving at the rate of 10 miles per hour, a man in the street would measure your new speed as 12 miles per hour, namely $10 + 2 = 12$ right? Likewise, if we consider light particles that move with speed c in the street, isn't it common sense that they should be observed moving 10 miles per hour faster when they are on that train? But shockingly, Einstein said no! According to him, the light particle emitted either on the street or on the train will still be observed to have the same speed c. This was hard to digest at first, but he was right.

The fact is that all inertial observers do indeed measure exactly the same speed of light c (in the absence of forces), whether from the vantage point of the street, or from the vantage point of the train, or any other vantage point in an *inertial* frame of reference.

What conclusion can we draw? What we didn't know before Einstein was that we were using the wrong rules for adding velocities! The simple rule of just adding them up, which seems to agree with everyday experience, is actually only an approximation to the correct rule. The exact formula, which Einstein discovered as part of his theory of relativity, does mathematically confirm that the speed of light is the same constant in all inertial frames.[1]

The basic fact that the speed of light is universal for all inertial observers leads to new relativistic rules that govern all motion, replacing those of Newton.

Some relativistic formulas are illustrated in Fig. 3.4 for which some further explanation can be found in a footnote.[2] It turns out that Newton's rules for computing velocity, momentum, and energy can be derived from these as approximations only for sluggish motion at speeds much smaller compared to the speed of light.

At sufficiently large speeds the relativistic rules make surprising predictions that are revolutionary and contrary to common experience in our daily lives. This is

---

[1]The correct relativistic rule is as follows. If the moving observer has speed v relative to a static observer, and if the moving one throws a particle with speed V in the same direction of motion, then the static observer will see the particle moving at a total speed $V_{\text{total}} = \frac{v+V}{1+\frac{vV}{c^2}}$. If the direction of motion of the particle is in the opposite direction, then V is replaced by $-V$. In this formula, if you replace V by c, then you will find $V_{\text{total}} = c$, and if you replace V by $-c$, then you will find $V_{\text{total}} = -c$. This formula illustrates how all inertial observers moving at any v see the same speed of light c. Furthermore, for any particle velocity V, if the moving observer himself has the speed of light, $v = c$, then again we find $V_{\text{total}} = \text{c}$, indicating that the static observer sees the other observer as well as any particle in his frame moving at the speed of light. Note that for speeds much smaller compared to c, we can neglect $\frac{vV}{c^2}$ compared to 1. Then the formula above reduces approximately to $V_{\text{total}} \approx \upsilon + V$, which agrees with experience in everyday life.

[2]According to relativistic rules, the relation between the momentum, the rest mass, and the energy of a freely moving particle is determined, as shown in Fig. 3.4, by a right triangle. The perpendicular sides are proportional to the momentum pc and rest mass mc^2, while the hypotenuse represents the energy $E = \sqrt{(pc)^2 + (mc^2)^2}$. For any momentum p, the speed of the particle is determined by the ratio of the vertical side to the hypotenuse, $\frac{v}{c} = \frac{pc}{E} = pc/\sqrt{(pc)^2 + (mc^2)^2}$. This gives the relativistic relation between rest mass, momentum, and velocity. From the figure one can see that the ratio $\frac{v}{c} = \frac{pc}{E}$ is always less than 1 as long as the horizontal side is not zero. Therefore, a massive particle $m \neq 0$ must always travel at speeds v less than c. Two limits of the triangle in Fig. 3.4 are of particular interest. Consider at first a very short horizontal side, when the hypotenuse approaches in length the vertical side. This represents a fast moving particle since v/c approaches 1, the maximum possible. For massless particles $m = 0$, such as the photon or light, the horizontal side collapses to zero. Then the hypotenuse has exactly the same length as the vertical side, and therefore $v = c$. So massless particles must always move at the speed of light $v = c$. Massless particles cannot stop no matter how small their momentum or energy is. Next, consider the opposite limit of a very short vertical side, when the hypotenuse approaches the length of the horizontal side. This represents a slow moving particle since v/c gets small. In the zero limit, the momentum is zero $p = 0$, so the particle is at rest $v = 0$. Then the energy E reduces to its rest energy $E = mc^2$, same as the horizontal side. When the horizontal side is much larger than the vertical side, which happens either when the rest energy mc^2 is very large, or the momentum pc is very small, then the formula for the energy E can be approximated by $E = mc^2 + \frac{p^2}{2m} +$ smaller terms, where mc^2 is dominant by far. The small part $\frac{p^2}{2m}$ is the kinetic energy (due to motion) of a slow particle, and it agrees with Newton's laws for kinetic energy. Similarly, in the same approximation, using E $\approx mc^2$ we obtain $v \approx \frac{p}{m}$ from the speed formula, which also agrees with Newton's relations $p = mv$ among momentum, mass, and velocity. So, Newton's formulas for energy and speed are valid only for the case of sluggish particles since they are only approximations to Einstein's relativistic formulas.

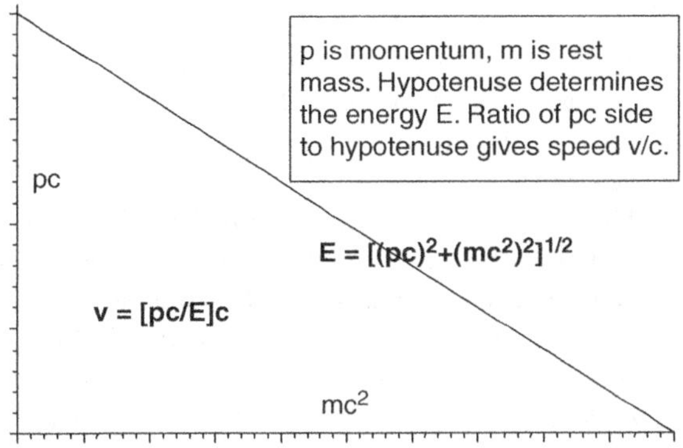

Fig. 3.4 Relativistic formulas for energy E and speed v are related to each other like the sides of a right triangle as explained in footnote 2

how Einstein first discovered that time and space are relative to the observer. One big surprise was that time measurements are not at all universal for all observers. Contrary to what was thought of before, according to static observers, their clock ticks faster as compared to an identical clock carried by a moving observer. So a static observer would measure a longer time interval in between two events since his watch will experience more ticks. This effect, which is called time dilation, gets bigger when the relative velocity between the two observers increases.

One of the consequences of time dilation is that in principle it is possible to travel to the future (but not the past) because, *according to stationary observers*, moving observers age more slowly than static ones. The cartoon in Fig. 3.5 illustrates this phenomenon: according to mission control on the Earth, the astronauts

Fig. 3.5 It's been only 20 years, why humans don't look like us anymore? Do you remember the time dilation factor that depends on speed as $\frac{1}{\sqrt{1-v^2/c^2}}$? (Credits, Brian Zaikowski.)

that are about to land were launched 50,000 years ago, but according to the astronauts, the whole trip took only 20 years. So, they are about to see the Earth 50,000 years later, rather than only 20 years later. In the meantime humans have evolved to look like the modern men and women rather than the little green creatures in the spaceship . . ..

The astronauts' 20-year time interval between launching and landing is dilated by a huge factor according to the Earth's observers. The dilation factor in this case is (50,000 years/20 years) = 2500. To obtain this much time dilation, the astronauts would have to travel at nearly the speed of light, $v = 299{,}999.9760$ km/s to be exact, according to the formula in Fig. 3.5. Of course, a speed close to the speed of light of 300,000 km/s is a huge exaggeration considering existing technology, but after all, this is only a cartoon. More realistically, manned space vehicles today can achieve speeds of about 12 km/s, a far cry from the speed of light. The time dilation factor at this speed is 1.0000000008, which means that the astronauts's watch will differ from mission control's by about a one millionth of a second per day. This is not noticeable in today's space travel since you need a trip of 1 million days to see a difference of about 1 s.

However, very high speeds close to the speed of light can be attained in particle accelerators for tiny particles such as electrons or protons. It is then possible to check the time dilation effect in the realm of particle physics. In the laboratory we can produce particles that have very short lifetimes, such as muons with a lifetime of 2.2×10^{-6} s. We can recreate the circumstances of Fig. 3.5 by substituting muons instead of people on the ship and at mission control. This is done by comparing the lifetime of muons that travel extremely fast in an accelerator to the lifetime of muons that remain static in the lab. The lab experiments show that traveling muons do indeed live a lot longer compared to the static ones, just as the traveling astronauts in the cartoon are returning back to the Earth centuries after their former relatives and friends are long gone. The dramatic effect of time dilation is then verified to be correct and exactly given by the formula in Fig. 3.5.

There is a famous *twin paradox* associated with time dilation. If one of the twins travels on a spaceship as in Fig. 3.5, then on his return he will find his earth-bound twin much older than himself. However, from the point of view of the traveling twin, he is static, while the other twin on the Earth appears to travel. So, when they meet, shouldn't the twin in the spaceship be older than the twin on the Earth? There appears to be a paradox since, according to this reasoning, each twin would think the other one is younger. In reality there is no paradox at all. The apparent contradiction arises by forgetting that the relativistic formulas discussed above apply *only to inertial observers*. The twin on the Earth is an inertial observer since no forces have been applied to it. But the twin on the spaceship is not an inertial observer. He experiences forces at take-off and landing as well as when he slows down to turn around and accelerate again to come back to the Earth. The reasoning above is incorrect because the traveling twin cannot be described with only the rules of motion of special relativity. When the correct rules, including forces, are taken into account, one finds that there is no paradox. The traveling twin is definitely the younger one.

We have illustrated how the time interval between two events depends on the speed of the observer. How about space? Again, static observers see space differently than moving observers. Static objects appear contracted to a moving observer. This is illustrated in Fig. 3.6. Space intervals between any two points *in the direction of motion* appear to be shorter according to a moving observer as compared to the measurement of the same interval by a stationary observer. Put another way, a moving stick, lying parallel to the direction of motion, appears shorter as compared to an identical stationary stick. This effect is called space contraction. Again, even for rocket speeds attainable today, the effect is too small to notice in daily life. However, for tiny particles moving at nearly the speed of light in particle accelerators, the effects of space contraction are confirmed to be precisely according to the formula in Fig. 3.6.

Here are a few related phenomena that would be noticeable at exaggerated speeds: A round disk moving at high speeds in front of you would look like an oval. At the speed of light the oval would shrink to a line segment. To a traveler moving at exactly the speed of light, the distance between the center of our Milky Way galaxy and the center of the Andromeda galaxy would look like zero.

Fig. 3.6 To an observer approaching close to the speed of light, the images on a flat picture contract in the direction of motion. For example the physicists in this picture at the Strings-95 conference appear to be tall and thin. The space contraction factor in the direction of motion is $(1 - (v/c)^2)^{1/2}$. The effect is more complicated in 3-dimensional space (rather than its picture) because the objects appear to also rotate

The special theory of relativity leads us to the following conclusion. Time and space are not absolute, but they are interrelated. They both depend on the observer. Each observer has his own space–time. Events that appear simultaneous in one space–time frame of reference are not simultaneous in moving frames of reference.

3.2 Equivalence Principle, Symmetry, and General Relativity

We have seen that some surprising effects arise when the same system is observed from certain unusual perspectives. Even more weird things happen for *non-inertial* observers that are under the influence of forces. The theory of general relativity, which describes the gravitational force, makes some rather surprising predictions.

For example, clocks should click faster when placed in weaker gravitational fields. So, people on the top floor of the Empire State building should age faster as compared to those on the first floor. Although this is a negligible amount of time on the scale of human lifetimes, it can be substantial for short-lived subatomic particles, so it can be checked. Experiments that checked this gravitational time dilation have been performed and agreed with the prediction.

Such bizarre effects become most pronounced close to the extremes. Close to the speed of light, or in very strong gravity, a watch appears to almost stop. Inside a black hole, time and space switch roles, etc. Such effects arise because observers (or clocks) are in relative motion, or under the influence of gravitational forces.

One may better appreciate what is going on geometrically by reviewing the space–time diagrams in Figs. 3.1 and 3.2. These illustrate space–time as seen by a static observer. However, from the perspective of a moving observer, whether inertial or non-inertial, those coordinate systems get distorted by the effects of time dilation, space contraction, and more complicated effects when there are forces. Hence the event history of Fig. 3.3 gets distorted into a very different curve from the point of view of non-static observers that examine the same history of events.

If one knows how to relate the coordinate systems of various observers in different conditions, then the distorted observations of non-static observers can be mapped onto the static coordinate system. In that way different observers can compare their observations from the same perspective.

Once we know how to relate different observers by coordinate transformations, we may ask, are there some quantities that are the same despite the different perspectives of the observers? If there are, then those quantities are called *invariants* under the coordinate transformations that relate those perspectives.

For example, just below Fig. 3.3 we mentioned that a different *static* observer located at a different origin of space–time sees the worldline of the particle in Fig. 3.3 displaced but maintaining the *same shape*. So, the shape of a worldline that describes the history through space–time is invariant under space–time *translations* of the coordinate system.

A more interesting invariant is the speed of light c for all inertial observers. That is, all those observers that may be moving, but not experiencing any acceleration, measure the same value of c. Similarly, there are other invariants under certain special coordinate transformations among certain special classes of observers.[3]

But most of all we would expect that the *fundamental* laws of nature must not depend on the coordinate system used by any observer. The laws should be invariant under all possible coordinate transformations, not only some special ones. This has to be implemented as a required symmetry property of the equations that describe the fundamental aspects of nature.

Einstein took a big step in realizing this concept for the first time mathematically when he proposed the *equivalence principle*. With this principle, plus some additional requirements, the only theory for gravity that was possible to construct turned out to be general relativity. This was amazing because a force such as gravity emerged only from requiring the principle that the laws of motion should be independent of all space–time perspectives of observers related to each other by general coordinate transformations.

Einstein's principle of symmetry under general coordinate transformations is an example of a *gauge symmetry*. The important aspect of a "gauge" symmetry is that it holds at each point of space or time independently than the symmetry at any other point of space or time. That is, each observer, located at his own space or time,

---

[3]In particular, one can define the concept of the *invariant relativistic distance* between any two events that any *inertial* observer would obtain when moving at any constant speed. This quantity is given by the following simple formula:

$$(\text{invariant distance})^2 = (\text{space interval})^2 - c^2(\text{time interval})^2.$$

We have explained before that the space or time intervals between any two events are different when measured by a static observer versus a moving one. Despite the fact that these intervals are different, for different *inertial* observers, the combination above gives the same numerical value for the invariant distance, for either the static or the moving inertial observer.

For example, if a particle of light travels from one point to another, we know that every observer must see the light signal moving at the speed of light. So for every inertial observer, the space interval, the time interval, and the speed must satisfy the standard formula (space interval) $= c \times$ (time interval). In that case, the formula above would say that, for any two events connected by a light signal, the invariant distance must be zero, according to the measurements of every inertial observer. Such events are said to be separated by a *light-like* invariant distance (i.e., 0). Similarly, when two space–time events yield a positive value in the formula above, they are said to be separated by a *space-like* distance, and when they yield a negative value they are said to be separated by a *time-like* distance.

For example, consider the two space–time events defined by you sitting in your chair looking at your watch at one instant, waiting a minute, and looking at it again. The space interval between the two events is zero since you did not move at all, while the time interval is 60 s. The formula above gives a negative result; this is consistent with the fact that the second event is separated from the first by a time-like invariant distance. Matter, energy, or people are able to travel between events separated by time-like or even light-like distances, but not between events separated by space-like distances. So, the universe we see today is only that part of the early universe that is connected to our present space–time by time-like or light-like signals. We cannot see those regions of the early universe that became space-like separated relative to our visible universe by moving faster than light during inflation.

can choose degrees of freedom independently than any other observer located at other points of space or time, as long as those degrees of freedom are transformed by the gauge symmetry. Gauge symmetries are so powerful that they restrict greatly the possible equations compatible with the symmetry, so much so that in some cases the equations are unique. General coordinate transformations are examples of gauge symmetries because each observer can choose his own coordinate system independently than other observers located elsewhere. The gauge symmetry of the equations of general relativity under general coordinate transformations is in fact the fundamental principle leading to a unique theory that correctly describes all gravitational phenomena.

How does gravity arise from this principle? When space–time is distorted, as illustrated in Fig. 3.7, the shortest path between two points is not the straight line that connects them. For example, on the surface of the Earth the shortest path between Los Angeles and Tel Aviv is along the great circle that is obtained by intersecting the globe with a plane that passes through those cities and the center of the globe. Similarly, in any curved space–time, freely moving particles follow such shortest paths called *geodesics*.

Free motion along geodesics can be viewed as being equivalent to curved paths as if created by applying a force on particles, so their worldlines deviate from a straight line. Thus, the effect of a force can equivalently be described as free motion in curved space–time, as in Fig. 3.7. Which kind of force would that be?

It turns out that general relativity correctly describes all known aspects of gravity as being due to the geometry of space–time. For example, a warped geometry, analogous to the distorted shape of a trampoline when a heavy sphere is placed in its center, is produced by the Sun. This distorts straight lines into geodesics, and it can be shown that motion along geodesics is equivalent to motion under the influence of the gravitational force of the Sun, such as the motion of planets. The surprise is that this applies also to light, so photons coming from another star should follow a

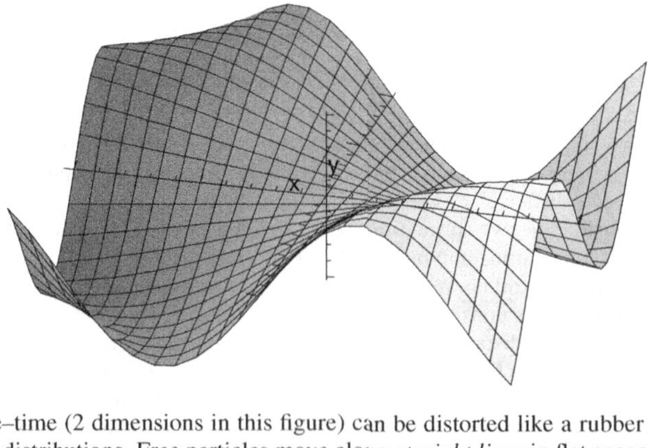

Fig. 3.7 Space–time (2 dimensions in this figure) can be distorted like a rubber sheet by nearby mass or energy distributions. Free particles move along *straight lines* in flat space–time, but along geodesics in curved space–time

curved trajectory that bends around the Sun. The predicted amount of bending was spectacularly confirmed by experiments soon after general relativity was proposed.

By contrast, Newton's law of universal gravity

$$\text{Force} = \frac{G \times \text{mass}_1 \times \text{mass}_2}{(\text{distance})^2}$$

would say that, since the mass of a photon is zero, it could not experience a gravitational force applied by the Sun; so a ray of light coming from another star would not be bent by the gravitational force of the Sun, which is contrary to observation. So, Newton's universal gravity had to be abandoned and replaced by the correct theory.

General relativity is a solution to an important conceptual problem. A crucial ingredient in special relativity was that in the vacuum of space nothing could travel faster than light. Einstein was bothered by Newton's law of universal gravity which allows two massive objects to communicate with each other instantaneously via the gravitational force. For example, according to Newton's laws, if you move a stone on the Earth, all masses in the rest of the universe are supposed to feel this change and react to it instantaneously. This is inconsistent with the finite velocity of light, which requires some time for signals to reach their destination. Einstein searched for a decade for a theory of gravity that would be consistent with the special theory of relativity, and finally, through the equivalence principle, he came up with the theory of general relativity which explained the fundamental origin of the force of gravity as due to gauge symmetry under general coordinate transformations. In this theory, gravitational effects which distort the gravitational force travel in free space as ripples that change the geometry of space–time, moving through it at the speed of light, consistently with the special theory of relativity.

While Newton's law of universal gravity works well as an approximation when the curvature of space–time is small, it is unable to describe the correct physics when the curvature is large. General relativity turned out to be the correct description of the gravitational force in all circumstances, from a more correct planetary motion, and the bending of light, to the gravitational forces in the interiors of stars, from the Big Bang to black holes, etc.

Is special or general relativity good for something practical? Yes, of course they are! Like everything else in fundamental science, eventually there are practical applications, and the apparently remote concepts of special or general relativity are no exception! Besides their indispensable roles at the scientific level in accelerator experiments, among the practical applications of these concepts, one should mention the global positioning system or GPS for short. For the GPS to work correctly, one must take into account the time dilation effects due to both special relativity and general relativity effects. The GPS has 24 moving satellites gravitationally bound to the Earth at 24,000 feet above the surface. To correctly compute your time and position on planet Earth on a continuous basis, the clocks on GPS satellites must work within 20–30 nanosecond errors. But the time dilation effects relative to clocks on the Earth are much larger than the error that can be tolerated. For satellites moving at about 3.9 km/s, according to special relativity the satellite clock would show a time difference of about −7000 nanosecond day. Furthermore, general relativity predicts +45,000 nanosecond time difference per day, relative to clocks on the Earth. If the clocks on the satellites are not adjusted continuously for these effects that amount to +45,000–7000 = +38,000 nanosecond, the same position on the Earth would be reported to shift somewhere else over time, and the GPS would be completely useless. The design of the GPS clocks has taken these effects of special and general relativity into account to get adjusted continuously and thus function correctly at the practical level.

The overwhelming evidence that accumulated for almost a century leads to the conclusion that general relativity is correct and must be part of any fundamental theory.

A major lesson learned from general relativity is that its success hinges on a very basic symmetry principle, namely the gauge symmetry under general coordinate transformations of space–time. This makes the fundamental laws independent of the coordinate systems used by any observer, including non-inertial observers. This is an example of a *gauge symmetry* principle that will be explained in the next section. The gauge symmetry principle was generalized later, following the leads of Maxwell and Einstein, into a comprehensive approach to describe not only the gravitational force but also the electromagnetic, strong, and weak forces as well.

The gauge symmetry principle is so powerful that the existence of these forces, as well as all of the detailed rules of how they govern all known aspects of nature, follow only from the requirement of the symmetry of the equations, and the degrees of freedom on which they act. The crucial part is to discover the correct symmetry and the correct degrees of freedom that govern nature.

According to current knowledge, the appropriate gauge symmetry that describes gravity is "general coordinate transformations" as discussed above, while the strong, weak, and electromagnetic forces are described by the Yang–Mills gauge symmetry based on a symmetry called $SU(3) \times SU(2) \times U(1)$, which is incorporated into the standard model of particles and forces.

When we discuss 2T-physics in later chapters we will see that its rules also emerge from a new gauge symmetry principle not noticed before. Let us then dabble in some notions of symmetry.

Chapter 4
Symmetry and Perspective

There are examples of symmetry in everyday life. Consider the symmetric artful designs in a palace, in a temple, or in a garden as in Fig. 4.1. The garden looks the same if you substitute one column for another, or if you interchange any two tiles of the same color on the floor. Under such transformations nothing seems to have changed even though we know that we have made some changes. Such transformations are examples of symmetries of the garden in this picture.

In Fig. 4.1 there is one garden, but there are many perspectives from which to view it. Different observers would take very different looking pictures depending on their standpoint. But if we are presented with a few different pictures of the same garden we find enough clues to integrate them together and reconstruct in our imagination much of the overall garden. Our senses are trained to do this. Let's

Fig. 4.1 Symmetry and perspective (Credits, Wikimedia Commons.)

I. Bars, J. Terning, Founding Editor: F. Nekoogar, *Extra Dimensions in Space and Time*, Multiversal Journeys, DOI 10.1007/978-0-387-77638-5_4,

concentrate on some aspects of the garden, such as symmetries and perspective, that our senses use to arrive at an overall picture.

For the sake of argument, let us assume that the symmetries of the garden are perfect, in particular that all tiles of the same color are identical and that every column is perfectly round and the same as another one, etc.

Based on the feeling of symmetry you can guess that the tiles that are far away, or the back sides of the columns that are not visible, must be identical to those that you see clearly as an observer from the perspective of this picture. Without the help of another observer, you can deduce that an observer located somewhere else, such as at the far end of the alley, will see the same kinds of tiles and columns. From your own measurements of sizes of tiles, circumferences of columns, etc., you can quantitatively tell exactly what the other observer's measurements will yield in another part of the alley. You are not deceived by the apparent small sizes of far away trees or other objects as they appear from the perspective of the photographer that took this picture. Rather, using the symmetry together with your daily experience on perspective, you can intuitively, and even quantitatively, estimate the sizes of the far away objects, such as trees, in proportion to other nearby objects in the garden.

In the same way, you can imagine symmetries of equations that describe nature. Such symmetries connect different perspectives to each other, thus permitting us to deduce many properties of the universe without having to put observers everywhere or under every possible imaginable circumstances. The symmetry transformations relate different potential observers to each other; their space–time positions, orientations, their speeds, or other degrees of freedom they would use in their equations to describe the universe, all get related to those of other potential observers. What we require is that the equations that describe the *basic laws of nature* look the same for every observer in their own frame of reference. Like the symmetric garden, the *fundamental laws of physics must be invariant* under the relevant symmetry transformations.

So, what kind of transformations must be symmetries of our basic equations? Part of the art in theoretical physics is to recognize in observations certain clues that relate to symmetries and from them devise the fundamental mathematical equations that are invariant under the appropriate symmetry transformations. Discovering what the symmetries of nature are is an ongoing pursuit as will be discussed in Section 7.2 for the case of two-time physics. But let us take a look at what is already established.

In what follows we will distinguish between two types of symmetries, namely *local* and *global*, whose meaning will be clarified. The aim is to explain that a local symmetry is much more powerful and that it is the basis of what is called *gauge symmetry* in fundamental physics.

From the success of general relativity and the standard model we know already that certain gauge symmetries are essential to describe nature. In particular, symmetries under general coordinate transformations and certain Yang–Mills gauge transformations are at the very foundation of these successful theories.

What is a *gauge* symmetry? To explain a gauge symmetry, we return to the tiles in Fig. 4.1. Let us assume that these tiles are all perfectly square, same size, without any designs on them. Now pick a single tile of any color and rotate it by 90°. You have actually made a change in the garden, but no observer can tell the difference in the appearance of the garden. This was a *local* rotation of an individual tile, performed by one observer at the location of the tile, while no observer located at other tiles made any changes. We can repeat this process for another tile somewhere else on this floor, and in fact we can do it for any number of tiles in various locations, without changing the garden at all. Local observers can make 90°, 180°, 270°, 360° rotations on individual tiles independently than one another without making any noticeable change in the garden. Such *local* transformations, which can be performed by observers at various locations independently from one another, that leave the garden invariant, are called *gauge* symmetries of the garden.

Now, let us contrast a *local gauge* symmetry with a global symmetry . This will shed more light on the meaning and power of a gauge symmetry. Suppose now we have a garden with tiles that have a picture of a flower on them such that they are all *oriented* toward you, the observer. If you rotate any *one* of them by 90° you obtain a new garden floor that looks noticeably different. So 90°, 180°, 270° local rotations of picture tiles are not symmetries of a garden that has such a floor (although 360° rotation is a local symmetry). However, suppose we ask all observers located on each tile to rotate them *all* by 90°. This is called a *global* transformation because it is performed *everywhere* in the garden. This global transformation actually changes the floor to a new one, but this is another symmetric state of the garden. The garden still has similar properties to the old one, for example, interchange of any two tiles of the same color is still a symmetry of the new garden.

The main contrast between a local and a global symmetry is that a global symmetry is associated with a transformation that is applied everywhere uniformly, while a local symmetry is associated with transformations that can be applied at any local point independently than any other. A local symmetry is also called a *gauge symmetry*.

What is the power of local symmetry over global symmetry? Local symmetry is much more demanding and puts stronger constraints. For example, in the construction of the garden suppose we require a local symmetry under 90° rotation of the tiles. This local symmetry forces us to use only uniform tiles. On the other hand if we require only a global symmetry, the tiles are permitted to have an oriented picture of any design, such as a flower. Finally if we do not require any symmetry at all, the tiles can be of any arbitrary design or shape in various parts of the garden. We see that the *local symmetry is the most restrictive*. It imposes the strongest constraints on the garden thus requiring that it is less varied or more uniform.

Now, instead of the garden let us substitute the equations that describe the fundamental laws of nature, and instead of the 90° rotation let us substitute certain special transformations of the degrees of freedom that appear in those equations. In the case of a fundamental theory of physics, to have a *global symmetry* means that the equations remain unchanged when *all observers everywhere* in the universe make the same transformation on the degrees of freedom that define the theory (whatever the transformation may be). On the other hand, the equations have a *local symmetry* if they don't change when *any single observer, anywhere or anytime in the universe*, makes a certain transformation of the degrees of freedom independently than any other observer located elsewhere in space–time. A local symmetry is much bigger and much more powerful because there is symmetry under an infinite number of transformations carried out by an infinite number of observers throughout the universe.

Which degrees of freedom are transformed by gauge symmetries and how? In the case of general coordinate transformations, the observer transforms his space–time coordinate system arbitrarily, not only into orthogonal coordinate systems as in Fig. 3.2, but he can deform and twist it around thus putting himself into the perspective of all possible space–time observers. This requirement is so powerful that it can be satisfied only by a unique "garden," namely the equations of general relativity.[1] The force of gravity must exist precisely because of this gauge symmetry.

In the case of Yang–Mills gauge symmetries, the degrees of freedom that are transformed are the gauge bosons, quarks, leptons, as well as any other form of fundamental matter needed to describe the universe, as described in Figs. 2.1, 2.3, 2.4 and 2.6 and the related discussion. The standard model, as a relativistic field theory, satisfies these gauge symmetry requirements for the symmetry group SU(3) × SU(2) × U(1) that was described in the text following Fig. 2.6. This also leads to a rather constrained theory, but not as a unique theory as general relativity. However, in this way the existence of the strong, weak, and electromagnetic forces emerge from the gauge symmetry itself, and all detailed interactions at the level of particle physics agree exquisitely with all observations so far.

Despite their tremendous beauty and experimental success, both of these gauge theories leave many questions unanswered. In both cases it is expected that the formalisms are valid only in a certain realm of energy or distance scales. The higher the energy the deeper distances can be probed, and so far available accelerators could probe the realm up to 10^{-18} m. At some point at distances deeper than this realm we expect to find that there is a more symmetric and a more unified beautiful theory that can answer the remaining questions.

We know many puzzles some of which will be listed later. Several approaches to resolve them continue to be studied vigorously. Deeper symmetries, acting on deeper matter structures such as strings rather than particles, and higher space–times beyond 3-space and 1-time dimensions, emerged as attractive avenues to find the essence behind the successes so far.

[1] Actually, symmetry considerations alone are insufficient to uniquely arrive at general relativity. An additional condition, that the equations should not involve more than two derivatives, is also needed to exclude other more complicated equations. The potential higher derivative modifications of general relativity actually occur as higher energy corrections that come from a more fundamental approach, such as string theory.

Chapter 5
Why Higher Space or Time Dimensions?

Measurable relics of the Big Bang include the observed background radiation at 2.7 Kelvin, the inhomogeneities in the energy distribution, the relative abundance of the chemical elements in our universe, and more. The Big Bang theory provides a quantitative explanation of what we see today, by relying on our theoretical understanding of the rest of the universe. The great success is that the prediction matches the experimental observation in quantitative detail. This is what gives us the confidence that the Big Bang and inflation theories are more or less correct.

We say more or less because our knowledge of the earliest stages of the Big Bang is still cloudy. Then the entire universe was very dense, very hot, under huge pressure. Matter in this early universe was moving at extremely high energies, much larger than the energies we can achieve today in accelerators for the tiniest of particles. Those early universe energies can probe distances that are as small as 10^{-35} m. However, our current understanding of both theory and experiment is firmly established only up to 10^{-18} m, which amounts to 1000 times smaller than the size of the proton. So we must ask, what were the correct laws of nature under the conditions of the early universe? Is it the laws encapsulated by the standard model for the subatomic world that includes quantum mechanics, and is there a role for general relativity?

Quantum mechanics should be applied to the tiny structures starting with the size of the atom, but general relativity must also play an important role to describe the space–time of the early curved and twisted universe. Indeed, general relativity is one of the essential tools for understanding the expansion of the universe starting soon after the Big Bang.

However, at the earliest times and extremely short distances close to the Big Bang, general relativity begins to give incurable mathematical problems and, in the context of quantum mechanics, turns into a non-predictive theory for physics. Building a quantum theory of gravity that includes general relativity has proven to be a challenge for the 21st century physicists. This is the big problem that led to string theory, and its extension, M-theory, as a resolution of how to marry quantum mechanics with general relativity.

If we believe that the string/M-theory approach is the only resolution for quantum gravity, noting that nobody knows a better one up to now, then we are led inevitably

I. Bars, J. Terning, Founding Editor: F. Nekoogar, *Extra Dimensions in Space and Time*, Multiversal Journeys, DOI 10.1007/978-0-387-77638-5_5,

to accept *extra space dimensions* beyond our common experience of 3-space and 1-time dimensions, as explained partially in the next section, and more fully in the next chapter on the role of the extra space dimensions in string theory.

From now on I will refer to the standard formulation of physics, which describes all of the above, as "1T-physics." All of the successful principles in 1T-physics are captured by general relativity and the standard model, while its possible less known deeper aspects take the forms of string theory and M-theory.

Unlike extra space dimensions, extra time dimensions present some conceptual physics problems that I will discuss below. Such problems could not be surmounted for many decades and for this reason it was difficult to imagine how more than 1 time dimension could work to describe physics.

However, it has been noted that supersymmetry structures in M-theory already give some clues of the possibility of an extra time dimension. The introduction of a new gauge symmetry in the context of two-time physics (2T-physics) in 1998 and the recent developments in this area, which includes the 2T-physics versions of the standard model and general relativity, overcame the conceptual issues with 2 time dimensions. The details will be given later in Chapter 7, but in order to provide some initial guidance for the reader, in this chapter I will provide a brief preview of how 2T-physics relates to 1T-physics. This will be captured and summarized by the shadows allegory in Section 5.4.

In the following sections of this chapter I will attempt to explain 2T-physics in simple terms, trying to avoid the predicament of the professor in Fig. 5.1. This is meant to give the reader a quick preview of 2T-physics and as such it is bound to skip details, leave some gaps in the reasoning, and raise some questions in the

Fig. 5.1 The main features of 2T-physics can be explained in simple terms, but not as clearly as the mathematical formulas (Credits, ScienceCartoonsPlus.com.)

reader's mind. A fuller explanation will be provided in the main discussion of 2T-physics in the coming chapters in this book. Here, I will use some analogies that can be understood intuitively in order to convey some of the fundamental ideas of 2T-physics. The reader should be warned that the analogies, which are never perfect, are meant to covey only some qualitative aspects of the theory; they should not be taken too far as they are not meant to faithfully capture the essential quantitative properties of the theory. In later sections I will discuss again the same concepts, and at times I will include some formulas in footnotes for readers that wish to understand the concepts in some mathematical detail.

It should be noted immediately that 2T-physics is not a speculative approach; it works as well as 1T-physics to describe the known aspects of nature, and in addition it goes beyond 1T-physics with its capacity to make predictions that 1T-physics is only able to verify but not able to predict in both the macroscopic and microscopic worlds.

2T-physics in 4-space and 2-time dimensions has successfully reproduced the usual 1T-physics as a "shadow" in 3-space and 1-time dimensions. The concept of "shadows" will be explained in the last section of this chapter. Taking together recent results for general relativity, the standard model, and supersymmetric 2T-field theory, these 2T theories correctly describe 3+1-dimensional nature directly in 4+2 dimensions. The successful theories of standard 1T-physics now have counterparts in 4+2 dimensions, thus providing a higher perspective on the significance of space and time, and lending a new outlook on unification of 1T-physics theories.

5.1 Extra Space Dimensions

As discussed in the next chapter in more detail, the notion of extra space dimensions first arose in the work of Kaluza and Klein, soon after Einstein's discovery of general relativity in 1916. But extra space dimensions were not a required ingredient until they resurfaced in the context of string theory in the 1970–1980s.

String theory was at first introduced by Y. Nambu and L. Susskind to understand strong interactions, but this role was better fulfilled by quantum chromo-dynamics which is now part of the standard model that was described above.

String theory was reinterpreted in the mid-1970s by J. Scherk and J.H. Schwarz as a theory of gravity, and later advanced to the frontline in the mid-1980s with the work of Green, Schwarz, Witten, Gross, and others, as a theory that potentially unifies all matter and forces. In that role string theory tries to answer simultaneously other puzzles of the universe.

String theory has provided the formalism for understanding quantum gravity together with the other forces without any mathematical inconsistencies. All forces and matter arise from tiny open or closed strings that split and join with each other. This concept unifies matter and forces.

The concepts in string theory have been taken one step further to include not only strings but also membranes, jellies, and higher dimensional structures called "*p*-branes." The "-branes" part of this nomenclature is analogous to mem-branes, while the "*p*" stands for an integer $p = 0, 1, 2, 3, \ldots$ that indicates the dimensionality of the object, such that 2-brane is the same as mem-brane, while the others stand for 0-brane = particle, 1-brane = string, etc. All of this is part of a larger, mathematically consistent theory that includes quantum gravity, called M-theory.

The precious mathematical consistency of string theory and M-theory leads to a surprising prediction. It requires extra dimensions beyond those experienced directly in our current observations of the visible universe comprising both the large and small structures. For example M-theory requires precisely 10-space and 1-time dimensions. We will discuss in a later chapter how to interpret the extra 7-dimensional space, and what observations would be a signal for their existence.

Furthermore, a scenario for how to connect this string picture, and its extensions, with the well-understood standard model and general relativity in 3+1 dimensions has been described and partially explored in the form of "string inspired models." The fact that such models can basically reproduce the details of the standard model allows us to be convinced that certain observed properties of matter, such as electric or magnetic charge and their generalizations that we call "flavors", "colors," and "families," can easily arise from the properties of matter as it moves in the extra dimensions.

Such concepts, which arise from what is known as "Kaluza–Klein compactifications" of space dimensions, will be a little better explained in the beginning of Chapter 7 after discussing concepts of space, time, and symmetries in the intervening chapters. The main point here is the fact that string/M-theory demands this scenario, and other related concepts of "large extra dimensions," in order to make contact between its 11 dimensions and observations in 3+1 dimensions.

However, the details of how such a scenario is *derived* from the fundamental dynamics of string theory or M-theory have remained elusive. Originally it was hoped that this scenario would one day be proven to be unique as a result of mathematical consistency in string/M-theory. Unfortunately the uniqueness aspects seem by now to have evaporated because the mathematical consistency of the known aspects of string/M-theory permits in the order of 10^{50} consistent scenarios, with most of them far from looking like our universe. Many string theorists have given up the hope of proving uniqueness and instead are content of using string/M-theory as the most consistent framework to model our universe in a way that is compatible with quantum gravity. Even within this less ambitious approach, the theory has proven to be mathematically very challenging, making it difficult to extract predictions that can be tested with our available technological or theoretical tools.

Does this mean it is time to give up on string/M-theory, as some critics suggest? Certainly not. At the very least, string/M-theory is the most reasonable framework that includes quantum gravity along with the standard model and classical general relativity. One should, however, realize that what has been so far explored as "string theory" is actually only part of a far more complete "M-theory" whose form remains largely unknown. We know some of its symmetry properties called "dualities" from which some information beyond string theory can be extracted. This is why M-theory doesn't have a full name; so M stands for "mysterious" or several other words (such as mother, marvelous, matrix) that start with the letter "M." Much more work is required to actually construct the theoretical basis of M-theory. When this job is accomplished we would be in a better position to make reliable predictions.

Today string theory or M-theory remains as part of the most viable and hopeful avenues to continue our exploration because conceptually there has not been a better proposal to solve the conundrum of quantum gravity. Extensive research continues to be done in this subject. This work proceeds with care because of the lack of verifiable predictions, but also with the trust that we are on the right track.

In my opinion there are still missing fundamental ingredients in M-theory as it is practiced today. At least part of what is missing is embodied in the fact that so far M-theory is thought to be a theory with only 1 time dimension. This already implies that M-theory, as discussed so far, cannot reproduce the dualities and hidden symmetries that are part of the secrets in Nature revealed by 2T-physics as I will describe in this book. However, M-theory has already given clues of 2 time dimensions. This clue has not been pursued by string theorists because of a lack of understanding of how to interpret a second time dimension. That gap has now been closed with the rapid recent progress in 2T-physics which has mostly been explored as a general approach quite independently than M-theory. To make further progress in fundamental theory one must find a way of combining the good aspects of M-theory and the powerful new tools of 2T-physics. I believe this will be a crucial ingredient in the theoretical basis of the eventual M-theory. Linking M-theory with 2T-physics is one of the goals in ongoing research in 2T-physics as will be described in the following sections.

5.2 A Matter of Perspective in 2T Space–time

Once the conceptual barrier of 3-space and 1-time has been breached by 11-dimensional M-theory, any curious person would immediately ask, why stop at only 1 time dimension, what about *extra time dimensions?*

It is just a stroke of the pen to naively extend space–time mathematically with extra time dimensions. To turn a space-like dimension to a time-like dimension requires only the insertion of a few minus signs in crucial places in mathematical formulas. But then, in the physical interpretation of the equations, one encounters physical inconsistencies in the form of negative probability called "ghosts," and other illogical problems associated with violations of causality (see Chapter 7). Those minus signs can actually cause calamities unless they are controlled by something else. Not knowing how to tame the minus signs, physicists generally stayed away from extra time dimensions.

However, as explained later in more detail, extra time dimensions had already made a subtle appearance in M-theory as I noted in 1995 [3] (explained later in footnote 1 of Chapter 7). The pursuit of extra time dimensions starting with this clue led to the discovery of two-time physics (2T-physics) as a completely consistent physical approach with 2 time dimensions.

2T-physics developed by finding the *fundamental solution* to the ghost problems, and related causality problems that arise when extra time dimensions are introduced. The solution was based on a gauge symmetry in phase space X^M, P_M, which arose as the new concept. 2T-physics has by now developed into a fairly general framework that is consistent with everything we know as encapsulated by the standard model and general relativity. Moreover, 2T-physics makes some new testable predictions and brings new perspectives for a higher unification in a space–time with 11-space and 2-time dimensions, as will be explained later.

It is only in the context of 2T-physics that a satisfactory resolution of the fundamental problems posed by an extra time dimension has been obtained. The difficulties with extra times, their resolution, and the surprising natural predictions that come with the resolution will be the topics of the coming sections.

A main lesson from 2T-physics is that there are physical phenomena in our own 3+1-dimensional space–time that 1T-physics is not equipped to capture, except accidentally in some instances. There exists hidden information in the form of higher dimensional symmetries and "dualities" (explained later), which is revealed by 2T-physics, but that is *systematically* missed in the 1T-physics approach. It turns out that 1T-physics can be characterized as being dependent on a certain perspective for observing the universe, thus missing on the bigger picture obtained in 2T-physics as a higher dimensional theory.

It took 13 centuries, from Ptolemy to Copernicus, for humans to be convinced the Sun and not the Earth is the center of the universe. The resolution in that case was just a matter of perspective, namely nearby regions of the universe look different if the observer sits on the Earth or sits on the Sun. But after recognizing this fact, we can easily reinterpret the observations as seen from the Earth. In this light the complexity of the theory advocated by Ptolemy can be understood as due to a transformation of perspective. So, his mathematical formulas for the motion of planets were not really wrong, but their complexity from the Earth's perspective obscured the correct interpretation of the observed motion of the planets.

I will explain in the coming sections that 1T-physics is formulated from the perspective of an observer in space–time and therefore has a bias in favor of treating [time, space] as more primary than [energy, momentum]. This bias is somewhat like viewing the universe from the perspective of the Earth and thinking that we are at the center of the universe. The analogous bias in 1T-physics forces a view of nature from a perspective which is not the most illuminating and obscures relationships between different 1T systems that actually exist but are not even suspected in 1T-physics. The formalism of 2T-physics is based on a larger symmetry that puts [time, space] and [energy, momentum] at the same level of importance in the description of the fundamental laws of motion both at the classical and quantum levels. This broader symmetry also requires a space–time that includes an extra space and an extra time dimension. This higher dimensional point of view eliminates the bias inherent in the 1T-physics formalism. In particular it teaches us that there are many lower dimensional 1T-physics perspectives from which one may view the same higher dimensional system, and that this provides new systematic information of many hidden relations among different 1T systems which are not evident in the usual 1T-physics formulation.

The bias of 1T-physics, and its dependence on different perspectives within a higher dimensional space, is explained intuitively by considering the shadows allegory discussed in Section 5.4 and in more detail in Chapter 7.

In this way we learn how to make a leap in our understanding of space–time and energy–momentum. This is somewhat similar to the way we learned in the past how to adjust the interpretation of our observations as we recognize the deeper laws that govern our universe. Newton was not wrong, but was merely describing the universe under some limited conditions. By now we know that the conditions for the validity of Newton's laws were that the sizes of structures had to be larger than the atom, their velocities had to be considerably smaller than the velocity of light, and the space–time in which they moved had to be nearly flat in the sense defined by Einstein. Under those conditions Newton's laws apply perfectly well even today, but only as an approximation of the deeper laws. Of course, because of the imposed limitations, the true nature of the universe is obscured, and its properties outside of the limits are not provided by Newton's theory.

It is in this sense that we need to understand the extra space and time dimensions proposed by the new theories, including 2T-physics and string/M-theory. The most successful theory today is the Standard Model of particles and forces that was described in the previous chapters. It is based on a spacetime of 3-space and 1-time dimensions. We know that it does not incorporate certain features, such as quantum gravity, dark matter, and dark energy, so we expect that a more complete theory should exist. The standard model will always continue to be the correct theory under certain conditions, in a similar way that Newton's laws are still valid.

In this light, it is important to emphasize that the new theories based on extra space or time dimensions are in complete agreement with the standard model and general relativity under the proper physical conditions, namely when we restrict the questions to the realm of 3-space and 1-time, and energy scales where the standard model or general relativity are valid, then the new theories are in full agreement with them. This agreement is a very first requirement for any new theory, and this test is successfully passed by appropriate scenarios of string theory or M-theory, as well as by 2T-physics. However, if the new theories are correct, the standard model and general relativity will be insufficient to describe the universe in the extra dimensions, or subtle phenomena in 3+1 dimensions.

5.3 Why Two Times

The fundamentally new basic feature of 2T-physics is a *gauge symmetry* called[1] Sp(2, *R*) which I introduced in 1998. The notions of "local gauge symmetry" as opposed to "global symmetry," and in particular the details of the "Sp(2, *R*) gauge symmetry" will be explained in detail in upcoming sections. All that needs to be understood at this stage is that "Sp(2, *R*)" is just a name that partially describes some of the mathematical features of the new symmetry.

The main new property of the Sp(2, *R*) gauge symmetry is that the basic laws of motion, as reformulated in 2T-physics, do not change when the primary quantities of motion, namely time, space, energy, momentum, are transformed into each other by the operations of Sp(2, *R*) at *any instant of any motion*.

Such a symmetry is not generally obeyed in 1T-physics. Although 1T-physics starts out with ingredients that have a similar symmetry in some parts of its formulation (example, symplectic symmetry of the Poisson brackets or quantum commutators), the symmetry is broken as soon as one concentrates on a specific system (example, a specific Hamiltonian). We will not regard this aspect of 1T-physics as a breaking of the local Sp(2, *R*) symmetry of the fundamental laws, but rather as an *apparent* rather than real breaking which is caused by adopting a *particular perspective*, namely the 1T-physics perspective, to describe the phenomena. There is a higher perspective, the 2T-physics perspective, that correctly describes the phenomena and does preserve the fundamental gauge symmetry Sp(2, *R*) .

Here is a brief introduction to the new symmetry. In 1T-physics we have become accustomed to symmetries of fundamental laws under transformations of [time $\leftrightarrow$ space (position)] among themselves, or [energy $\leftrightarrow$ momentum] among themselves, as championed by Einstein. The new feature discovered in 2T-physics, that will be discussed in great detail later, is that the fundamental laws of motion have a larger

---

[1] A little mathematical background on Sp(2, *R*) as a symmetry group: The mathematical language of symmetries in physics is group theory. In the 19th century, the French mathematician Elie Cartan classified all the so-called Lie groups, and the corresponding Lie algebras, into seven "simple" classes that he called *A, B, C, D, E, F, G*. There are an infinite number of Lie groups in the following classes A_n, B_n, C_n, D_n, each labeled by an integer n that takes values $n = 1, 2, 3, \ldots, \infty$. The *E, F, G* classes are called "exceptional" and there are only a few of them labeled as E_6, E_7, E_8, F_4, G_2. In terms of Cartan's classification, the Sp(2, *R*) Lie group relevant for 2T-physics is identified as C_1. But in this classification $A_1 = B_1 = C_1$ are all the same as long as they are all taken in their *real* forms (i.e., only real numbers are involved in all transformations). Hence another name for Sp(2, *R*) $= C_1$ is SL(2, *R*) $= A_1$, and both of them are the same as Spin(1, 2) $= B_1$. In any case, in Cartan's classification, Sp(2, *R*) is the smallest possible "simple Lie group." The acronym Sp(2, *R*) is developed from the large characters in the following description of a specific group of transformations: "Symplectic transformations in 2 dimensions as applied on Real numbers." In the present application of Sp(2, *R*) in 2T-physics, the "2" does not stand for two dimensions, but rather for the 2 symbols *X* and *P* which are the covariant position X^M and contravariant momentum P_M, in every direction of space–time labeled by *M*, including the extra dimensions. Symplectic transformations have the property of leaving invariant the Poisson brackets in classical mechanics or the quantum commutators in quantum mechanics. This is the main reason for why symplectic transformations applied in phase space X^M, P_M is relevant in 2T-physics.

set of symmetries when *space–time* is transformed into *momentum–energy*, and vice versa, as sketched by the vertical double arrows in the diagram below:

time $\leftrightarrow$	position (space)
$\updownarrow$	$\updownarrow$
energy $\leftrightarrow$	momentum

In this diagram the vertical transformations (as well as more complicated phase space transformations that do not fit in this diagram) are implemented by the Sp(2, R) gauge symmetry *at any instant of motion.*

Guided by this new symmetry, the laws of both classical and quantum mechanics must be rewritten in a space–time of 2-time and 4-space dimensions in a form that displays the symmetry at *any instant of any motion.*

It turns out that 2 time dimensions is not an input but is a consequence of the Sp(2, R) symmetry. It will be clarified later that two times is actually a requirement without which the Sp(2, R) gauge symmetry cannot be realized in the fundamental laws of motion. Hence the formalism of 2T-physics, and in particular the 2-time coordinates, follow uniquely from the principles of the Sp(2, R) gauge symmetry. Furthermore, more than two times cannot be allowed because the available gauge symmetry cannot remove the ghosts and the causality violations. So, Sp(2, R) gauge symmetry requires exactly 2-time dimensions, no less and no more.

More generally, the Sp(2, R) gauge symmetry is compatible with d space and 2-time dimensions. There is no restriction on the number of space dimensions, so d can take any integer value, $d = 1, 2, 3, \ldots$, where the permitted extra space dimensions beyond the usual 3 dimensions may be similar to the extra space dimensions of string theory.

Why have such an Sp(2, R) symmetry in the first place? The answer to this question is because it restores [time, position] and [energy, momentum] as being equally important in the fundamental formulation of motion, for *all motion* both at the classical and quantum levels. Just on the basis of aesthetics, this is much more appealing than 1T-physics which lacks this symmetry.

But there is much more than aesthetics. The resulting 2T-physics theory goes beyond 1T-physics in describing phenomena in nature and shows that the 1T-physics is actually incomplete. The 2T point of view, together with the underlying Sp(2, R) gauge symmetry, is *necessary* to obtain a more complete and more unified description of observable phenomena.

As described in the previous section, 1T-physics is formulated from the perspective of an observer in space–time and therefore has a bias in favor of treating [time, space] as more primary than [energy, momentum]. The Sp(2, R) gauge symmetry removes the bias for a space–time perspective that creeps inadvertently into the formulation of 1T-physics. The limitations created by the bias in 1T-physics become clear from the new results provided by 2T-physics, which include new dualities, hidden symmetries, and unification of 1T systems predicted by 2T-physics for observers accustomed to the tools of 1T-physics. Examples of these predictions and the new properties that they reveal will be discussed in later sections.

With the extra dimensions, the more symmetric fundamental laws capture more information and provide a unification of 1T-physics that was hard to notice by relying only on the usual formulation of physics. An observer in 3+1 dimensions is at a disadvantage because he/she has only the usual 3-space and 1-time to grasp motion that takes place in 4+2 dimensions. Observers in 3+1 dimensions can only see shadows of the higher dimensional system, but can detect the presence of an extra time and an extra space by analyzing and comparing different 1T-physics systems as described below.

The effect of the Sp(2, *R*) gauge symmetry is that motion in 4+2 dimensions does not include all possible motions, namely only a subset of all possible motions that is symmetric under Sp(2, *R*) gauge transformations is permitted. This symmetric subset is found to be equivalent to motion in 3+1 dimensions and is best explained by the shadows analogy described below. The important thing about the analogy is the *many* shadows of the same object. These many shadows provide *many perspectives* of the higher dimensional systems, and it is through the many perspectives that an observer in 3+1 dimensions can detect the presence of the extra dimensions.

5.4 The Shadows Allegory

To better explain this, we will first use an allegory. Consider the relation between an object moving in a 3-dimensional room and its many shadows on walls created by shining light on it from different perspectives. Figure 5.2 illustrates some of the possible shadows, but shadows of intermediate shapes of a moving object are also included in our considerations. To observers that live *only on the 2 dimensional walls*, the different shadows appear like different "beasts" making various motions as if unrelated to each other. However, with great diligence and a lot of research these 2 dimensional observers can in principle discover lots of relationships among the shadows and their motions, since after all every shadow comes from the same object in 3 dimensions. All the possible relations that can be discovered in this way are of course predicted and unified directly in the room. When the 2-dimensional observers finally put it al together, ultimately they can reconstruct the unique 3-dimensional object and its motions. All of this is easily grasped by an observer that has the privilege of being in the 3-dimensional room. Evidently, the perspective of being in the room is much more powerful than the limited perspective of an observer confined on the wall.

In a similar way, it will be clarified later in detail that we are observers like those on the walls that are "stuck" in 3+1 dimensions. We do not have the privilege of being in the "room" in 4+2 dimensions. 1T-physics is a formulation of the phenomena on the 3+1-dimensional "wall." By contrast, 2T-physics provides the means to capture the perspective of the observer in the 4+2-dimensional "room."

In 3+1 dimensions we observe many facets of the universe and have managed to formulate the basic laws in the form of the standard model and general relativity, as

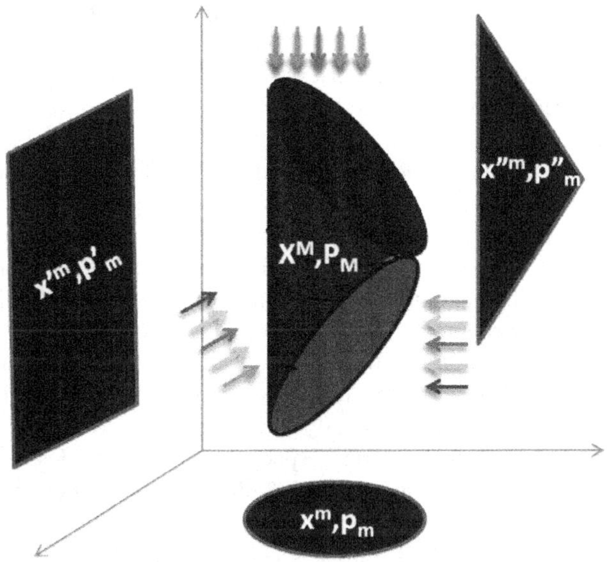

Fig. 5.2 Shadows of the same object appear different but must be related. In this allegory, the object in the room represents phase space (X^M, P_M), which consists of covariant space–time X^M and contravariant momentum–energy P_M in d+2 dimensions, while the shadows represent the many possible corresponding phase spaces (x^m,p_m), (x'^m,p'_m), (x''^m,p''_m) in $(d-1)+1$ dimensions

described in the previous sections. According to 2T-physics, all of this is analogous to the observers and the shadows on the walls. Observers in 3+1 dimensions see only many shadows of the same unique system in 4+2 space–time. Because of the unfavorable perspective, observers in the lower space–time have a hard time noticing the relations among the shadows that come from the higher space–time. The mathematical formalism of 1T-physics is not adequate to capture the relations among the shadows, since from the perspective of 1T-physics each shadow is described separately as a system of equations in 3+1 dimensions.

However, the 2T-physics formalism in 4+2 dimensions provides the higher perspective to automatically yield many hidden relations and symmetries in 3+1 dimensions that relate to the extra dimensions. These are predictions from 2T-physics that observers in 3+1 dimensions must find, and therefore verifying the existence of hidden symmetries and hidden relations in 3+1 space–time, as well as certain additional properties, provide some of the tests of 2T-physics. Such tests have already been verified in simple systems (see, e.g., Fig. 7.7) as discussed later in this book, so we know that the predictions of 2T-physics are valid in the real world.

In this way 2T-physics continues to make predictions and provide tests for observers trained in 1T-physics in one less space and one less time dimensions. According to 2T-physics, the fundamental rules of physics, as we have observed them so far in the macroscopic or microscopic worlds, have their origins in 4-space and 2-time dimensions.[2]

How about our most successful theories? As 3+1-dimensional observers, we have organized our current physics knowledge into the successful theories of general relativity and the standard model in the language of 1T-physics. It has been shown that in fact these arise from 2T-physics as particular shadows of their respective parents as 2T-field theories in 4 + 2-dimensions [38, 43]. Hence 2T-physics correctly describes 3+1-dimensional nature directly in 4+2 dimensions. The new perspective of 4+2 space–time leads to new testable predictions at all distance scales, large and small, in 3+1 space–time, and provides some new guidance at the fundamental level of particle physics and cosmology, as discussed later.

2T-physics is more generally formulated in a space–time with any number of space dimensions but only 2 time dimensions, not less and not more. This is because its symmetry is mathematically and physically consistent only for two times. Normally it casts shadows on a space–time with one less space and one less time dimensions. So 2T-physics in 4+2 dimensions creates shadows in 3+1 dimensions.

Similarly, if M-theory in 10+1 dimensions is regarded as a shadow, then the parent 2T-physics theory would be in 11+2 dimensions. This 13-dimensional theory, which is yet to be constructed, would cast many 11-dimensional shadows with lots of relations among them. In the mid-1990s I discussed some symmetry properties of such a 13-dimensional theory under the name of S-theory [5–9]. Those were intimately related to duality symmetries of M-theory which will be outlined later (see Fig. 10.1). So it is likely that the known dualities among the different corners of M-theory, which are still a mystery today, would emerge as part of the relations among the shadows created by the space–time perspectives in 11+2-dimensional 2T-physics which is yet to be constructed.

We are currently at a stage of still searching for the deeper formulation of the forces and matter in the universe and their unification. The unifying higher perspective of 2T-physics is expected to play a fundamental role for this quest.

[2] Students of philosophy may find some similarities between the shadows analogy used to explain 2T-physics and Plato's allegory of the cave (see, e.g., http://en.wikipedia.org/wiki/Allegory_of_the_cave). Indeed some familiarity with Plato's ideas may be helpful in understanding one aspect of the shadows, namely the question of how observation is related to reality? But there are actually important differences between 2T-physics and Plato's ideas. In particular, in Plato's allegory, there is a one-to-one correspondence between a real "form" and the corresponding shadow. But in 2T-physics there are many shadows of the same 4+2 events or history that are interpreted as different from each other by observers in 3+1 dimensions. The crucial point in 2T-physics is that it predicts hidden relations between apparently distinct 3+1 systems, thus making testable predictions that correctly fit nature. 2T-physics *unifies* quantitatively, not just descriptively, distinct 1T-physics systems into one 4+2-dimensional system.

Chapter 6
The Role of Extra Space Dimensions in String Theory

In the Einstein era of general relativity the only other force known to exist beyond gravity was electromagnetism which was already successfully described by Maxwell's equations. There was a desire to try to understand if these theories had a common origin. Einstein himself had the goal of unifying Maxwell's theory with general relativity.

The property of matter that responds to the gravitational force is its mass, or more generally, its energy, and its momentum. These properties are closely connected to its behavior in space–time. On the other hand the property of matter that responds to the electromagnetic force is its electric charge and its magnetic moment. These properties of matter go beyond its behavior in space–time.

It was quite bewildering how to unite space–time properties of matter with its electric and magnetic properties. Without finding a similar origin for such properties of matter the dream of uniting electromagnetic and gravitational forces could not be realized.

The concept of extra space dimensions, which originated soon after Einsteins theory of general relativity, does precisely just that!

Kaluza (1919) and Klein (1926) considered general relativity in 4-space plus 1-time dimensions. Their new fifth dimension was just a small circle of radius R characteristic of the size of this extra space. To explain why this extra space is not directly apparent to us, the size of the radius R was assumed to be extremely small to be visible by current technology. When matter performs periodic motion around a circle, it cannot have arbitrary values of momentum, but rather its momentum can only take certain discrete values given by an integer multiple of (h/R), where h is Planck's constant underlying quantum mechanics. So the fifth component of momentum, which describes motion in this tiny circle of Kaluza and Klein, can only have the quantized values 0, $\pm(h/R)$, $\pm 2(h/R)$, $\pm 3(h/R)$, etc. Accordingly, in this kind of 5-dimensional space, matter has ordinary momentum–energy in the usual 4 dimensions and also a quantized amount of momentum along the fifth dimension. Kaluza and Klein interpreted this quantized fifth component of momentum as the electric charge of matter, so that h/R is identified with the smallest electric charge e, namely that of the electron. The larger the momentum of matter along the fifth dimension (i.e., the faster it moves along the circle), the larger its electric charge. In

I. Bars, J. Terning, Founding Editor: F. Nekoogar, *Extra Dimensions in Space and Time*, Multiversal Journeys, DOI 10.1007/978-0-387-77638-5_6,

this way, electric charge can be regarded as being a space–time property of matter in the larger space–time.

Kaluza and Klein found also a candidate for the electromagnetic field as part of the theory in the extra dimensions. The extension of the gravitational field into 5 dimensions could be separated into two components. One part of the gravitational field related to the usual Einstein equations of general relativity is 3+1-dimensional space–time. The second part of the gravitational field, which contained the information about the fifth dimension, behaved just like the electromagnetic field in Maxwell's theory of electromagnetism in all of its mathematical details.

So Kaluza and Klein did manage to unify the gravitational and the electromagnetic forces in the form of general relativity in 4+1 dimensions, albeit with some conditions on the geometrical shape and size of space in the extra dimension.

The new insight that came along with this unification, unless it is only a mathematical trick, was the suggestion that *at every ordinary space–time point* there should exist 1 extra space dimension in the form of a tiny circle. But the circle had to be so tiny that it should be hard to notice or travel inside it.

Put another way, our current instruments are too bulky to probe the tiny circle. But a tiny "Christopher Columbus," such as a very fast moving electron, can in principle insert itself into the circle, explore the small dimension, and bring the information back to us. So, with sufficient amount of energy it is possible in principle to travel around the extra dimension of Kaluza and Klein. In this way it is possible to imagine ways in which the concept of the extra dimension can be tested. This idea is illustrated in Fig. 6.1

Like many other new attractive ideas at every stage of the development of physics, except for a core of physicists, the majority often seem to resist new ideas

Fig. 6.1 The large acrobat senses only 1 dimension, but the small flea can explore 2 dimensions by circling around the rope (Credits, Particle Data Group.)

even when nothing is wrong. To be sure, not every speculative idea that seems attractive at one point or another will survive the test of time. The broader acceptance requires additional compelling reasons in favor of the new concepts. Whether a fifth dimension exists or not, historically it did not make any difference in how the successful equations of general relativity or electromagnetism were applied in physics for the following 60 years after the work of Kaluza and Klein. So, the concept of extra dimensions faded away and came back several times. Finally extra space dimensions became part of the actively pursued frontline theories when they became essential to resolve a conceptual crisis.

The crisis emerged from the increasing tension between general relativity and quantum mechanics. These are the two most successful theories of the 20th century. The problem became apparent by realizing that gravity behaved somewhat differently than the other familiar forces at very small distances where quantum mechanics was the rule of law. This prevented gravity from being unified with electromagnetic, strong, and weak interactions, because the other forces were successfully unified and correctly described at the *quantum level* in the context of the standard model . Any attempt to include gravity as part of this unification of forces came into violent conflict with quantum mechanics. If there is real unification of forces then it had to happen at the quantum level.

Why unification, is it necessary? In particle physics, which is well in the realm of quantum mechanics, gravitational forces play no role whatsoever in determining the outcome of current experiments. At the level of 1000 times smaller than the size of a proton, the strength of the gravitational force on quarks or electrons is about 10^{41} times smaller compared to the electromagnetic force (see Figs. 2.3 and Fig. 9.2 and footnote 1). These forces are far from playing comparable roles at those quantum levels, so what is unification good for even if it is true?

The point is that the current level of experiment is not the guide for this issue. We need to look deeper, to the era of the Big Bang when the strengths of all known forces can be *estimated to be indeed comparable*. At that early time it makes sense to ask whether the forces have a common origin? If one wishes to answer it positively, there is no escape from confronting the issue of quantum gravity. Without reaching a conclusion on what quantum gravity is all about there is no way to be convinced of what precisely happened at the Big Bang.

Decades of searches for alternative theories that could resolve the issue of quantum gravity did not yield good enough candidates until the 1980s when string theory was interpreted as a theory of gravity by Joel Scherk and John Schwarz. According to string theory, all forms of fundamental matter or energy we know today at about 10^{-18} m (see Figs. 2.1, 2.2, 2.3, 2.4, 2.5, 2.6), namely quarks, leptons, gauge bosons, are all really tiny strings, so tiny that the strings appear to be point-like from our *current perspective*. It is only from the perspective of much smaller scales of about 10^{-35} m, which occur at the Big Bang, that the string-like behavior of all matter and energy finally would become apparent, as in Fig. 6.2.

String theory relies on the assumption that the structure of matter is sufficiently simple so that there are no new layers of matter between the currently known regime of about 10^{-18} m and the much smaller Big Bang regime of 10^{-35} m. This

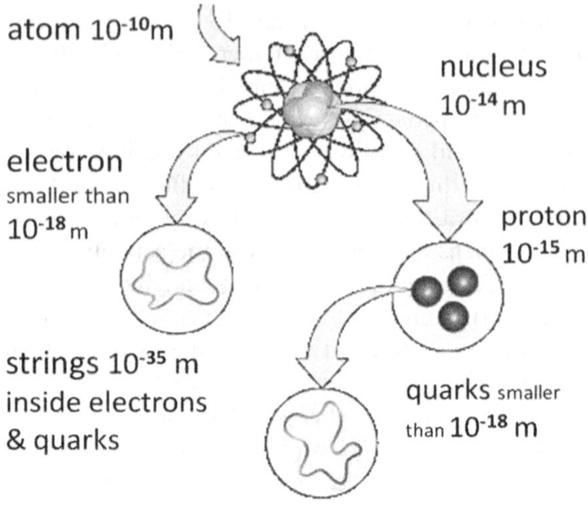

Fig. 6.2 Strings as the smallest constituents of matter (Credits, Particle Data Group.)

assumption may be a big leap of faith, but it is the simplest possibility which is still consistent with all known data, including the data that explain the entire history of the universe since the Big Bang. Other proposed alternatives of more complicated layers in the structure of matter are not impossible, but have their own problems and are less justified according to current understanding.

Crucial properties of string theory that resolve the quantum gravity issue include the fact that elementary matter is extended, string-like rather than being point-like, and that there must be a stronger symmetry, called *supersymmetry*, that makes bosons and fermions indistinguishable at the fundamental level. Gravity is not inserted in string theory as an adjunct force. It emerges even without asking for it from the only possible interaction, which is the process of string joining and splitting of *closed* strings in the shape of loops. The other forces also come about from the same origin, but involve the joining or splitting of open strings, in the shape of worms. These features provide a mathematical framework in which all forces as well as all matter can be unified with each other, while they are all consistent with the rules of quantum mechanics.

The scenario of string theory is quite attractive. Nearly 25 years of research since the mid-1980s has yielded a wealth of new insights, new ways of thinking about the fundamentals, and new computational techniques that can be applied to a variety of problems in string theory, gravity, particle physics, as well as in other fields of physics or mathematics. However, the details that connect string theory to the standard model, which was the original motivation for this scenario, have not yet been developed to a convincing level.

The basic complaint of critics of string theory is that the remaining mysteries of the standard model seem to be just as unanswered now as they were in the 1980s. String theory that seemed initially poised to resolve them has not yet delivered a convincing answer.

It is not clear that much of the criticism is warranted since no better alternatives than string theory have emerged to solve the difficult issues it addressed. The fact that the theory turned out to be more difficult than originally envisioned is not a fundamental failure in itself. Many adherents of string theory continue to get encouraged by the incessant technical developments and expect that eventually the standard model will be explained.

By now it has become evident that the fundamentals of the theory lie deeper than strings. Higher extended objects, membranes, jellies, and so on, generally called *D-branes* and *p-branes*, must interact with strings. A more comprehensive theory that takes all of that into account is envisioned. It goes under the somewhat mysterious name of *M-theory,* whose basic equations are not known, but only some of its properties can be identified at present. Constructing and solving M-theory is a major goal. But despite many proposals, the right idea does not seem to have emerged yet.

Now we return to the unification ideas through extra space dimensions. As far as a viable theory of quantum gravity is concerned, string theory and its extension into M-theory remain as the only known conceptual solutions to the issue.[1] The other fascinating aspects of this theory can be regarded as bonuses that remain to find their proper roles. To attain the required mathematical consistency at the quantum level, string theory must be constructed in space–time that include as many as 6 extra space dimensions, while M-theory must have 7 extra space dimensions.

This is regarded as one of the immutable predictions of this theory. As shown in Figs. 6.3 and 6.4, we expect then events and event histories in a total of 11 dimensions. So one must ask *where is this extra space and what role does it play?*

One immediate answer is provided by the Kaluza–Klein idea of small curled-up dimensions. This comes to rescue string theory from the embarrassment of extra dimensions, by turning them into a virtue rather than a problem. The components of the momenta in the extra curled-up dimensions play similar roles to the electric charge in the Kaluza-Klein proposal.

Explaining the patterns of generalized color and flavor charges of quarks, leptons, and gauge bosons (see Figs. 2.3, 2.4, and 2.6) is tantamount to understanding

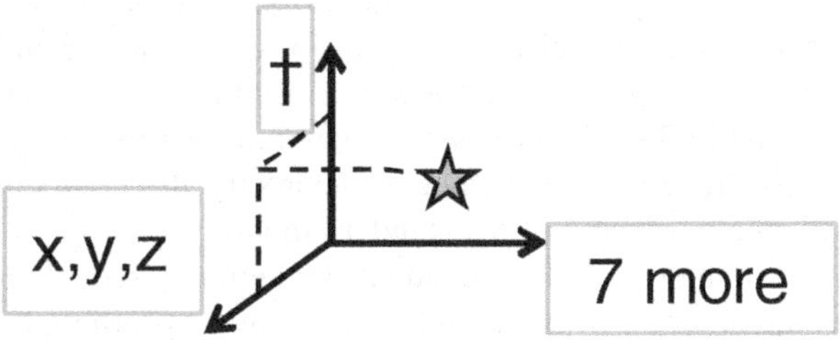

Fig. 6.3 M-theory: 7 more space dimensions

[1] Attempts to quantizing ordinary general relativity, such as *loop quantum gravity,* by rewriting Einsteins theory in terms of more elegant variables, are far from providing the resolution of the original quantization problems.

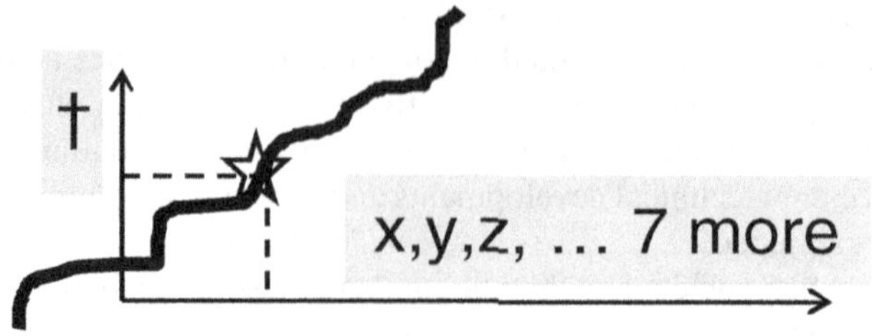

Fig. 6.4 History in 11 dimensions

the details of all their interactions via the principle of gauge symmetry. The known charges come as certain patterns defined by the symmetry group SU(3) × SU(2) × U(1), and they repeat in three families of quarks and leptons. Why these patterns, and especially why three repetitions, have been among the most basic mysteries of the standard model?

It is extremely interesting that it is possible to explain mathematically that those color and flavor patterns can arise from certain special geometries in the extra 6 or 7 dimensions by generalizing the Kaluza–Klein idea that electric charge can emerge from motion in the extra dimension. Connecting the SU(3) × U(2) × U(1), symmetry patterns of each family, as well as its *repetition in three families*, to a specific underlying geometry in the extra dimensions called Calabi–Yau manifolds is one of the fundamental conceptual successes of the approach with extra space dimensions.

If this is not just a mathematical accident, but reflects reality in the sense that additional verifiable consequences also would follow, then one could say that we have already observed some of the effects of the extra dimensions by the very existence of the quarks and lepton families in Fig. 2.6

How did the universe end-up with its current state in which certain curled-up geometries in the extra dimensions exist simultaneously with an expansion in the familiar 3+1 space–time? Here is where we do not have yet a complete theory. We hope that there is a right theory that will be able to explain a scenario similar to the following one, which remains still as a possibility within string or M-theory.

In the beginning all 11 dimensions existed on an equal footing, and matter or energy could travel in all dimensions as indicated in Figs. 6.3 and 6.4. Soon after the Big Bang, a phase transition occurred, after which the universe continued to expand in 3+1-dimensional space–time, but in the remaining 7-dimensional space, it got frozen to some specific geometry that explains the SU(3) × U(2) × U(1) symmetry content of the standard model.

Among the challenges in string theory is to explain why 3+1 dimensions, rather some other number expanded, and why a specific geometry, rather some other one, ruled in the extra dimensions?

In summary, some of the effects of extra *space* dimensions can be interpreted as being certain properties of known matter, such as electric charge, and similar other properties of quarks and leptons, that are not predicted by the standard model. While this attractive possibility could be taken as evidence for extra dimensions, this alone does not prove extra space dimensions. It is like the tip of an iceberg. To be convinced, we need to observe a lot more of the predicted details, of what would happen when we send in probes that can penetrate deeper, and see at least some aspects of what is inside the extra space.

Alas, in most models of the Kaluza–Klein-type scenario, the tiny geometry is too small to penetrate with any of the foreseeable accelerator energies. There may be possible effects of the tiny geometry in cosmological scenarios, dating back to the era when a lot more energetic particles capable of penetrating the small geometry existed. This has not been well explored, first because the fundamental theory underlying string theory has not yet been constructed, second because it is a difficult computation in the context of strings propagating in curved spaces.

There are also more general scenarios for extra space dimensions that are called *brane-world* scenarios following proposals by Randall and Sundrum. This will be discussed by Prof. Terning in the second part of this book. In this approach not all of the extra dimensions have to be curled up into tiny geometries, some of them could be large. These scenarios present alternative overall views of the universe that, under certain optimistic assumptions, can lead to observable effects at the upcoming Large Hadron Collider at CERN. Since this approach will be discussed in another chapter in this book, I will not go into more details here.

Chapter 7
Two-Time Physics

After the Kaluza–Klein scenario found a home in string theory, physicists became accustomed to the concept of extra space dimensions. Their presence was essential for the mathematical consistency of the theory. Adding extra dimensions of space now seemed natural and even necessary, but physicists shied away from adding also extra dimensions of time. Why?

Extra time dimensions had actually been quite discouraging to a lot of theorists because extra times bring additional problems that nobody knew how to resolve for many decades.

After clarifying the problems with the extra time dimensions, I will explain that the key to their resolution has to be a new gauge symmetry. The crucial question is then how to find the appropriate gauge symmetry that agrees with nature.

The arena for the new symmetry turns out to be *phase space*. This is the larger set of variables of motion that include not only coordinates in space–time but also coordinates in momentum–energy. Previous gauge symmetries acted locally in space–time, leading to physically consistent theories with only *one time*, no less and no more. The new gauge symmetry transforms all of phase space, namely space–time–momentum–energy, and it acts on these degrees of freedom at any instant for any motion. Requiring such a symmetry turns out to be physically and mathematically consistent only with *two times, no less and no more.*

Remarkably, the new gauge symmetry, called Sp(2, R), resolves all the problems and uniquely leads to the formalism of 2T-physics. The presence of two times requires a re-examination of the fundamentals of mechanics and the role of time. The new formalism captures 1T-physics as part of its structure and goes beyond by providing a new path to a new type of unification that 1T-physics on its own could not accomplish.

In the following sections the motivations, the path of discovery, the nature of the Sp(2, R) symmetry, and its predictions will be described.

7.1 Historical Path to the Sp(2, R) Symmetry

Why don't we find much discussion of more time-like dimensions in the physics literature? It is because extra times are risky, dangerous, and scary to many

I. Bars, J. Terning, Founding Editor: F. Nekoogar, *Extra Dimensions in Space and Time*, Multiversal Journeys, DOI 10.1007/978-0-387-77638-5_7,

physicists. Previous attempts to naively add extra time dimensions to existing theoretical frameworks always ended up in disasters.

There are two main pitfalls that are commonly expected with extra time dimensions: ghosts and causality violation.

- Ghosts: These are states of a quantum mechanical system predicted to occur with negative probability. Probability must be positive, so this is nonsense. A physically sensible theory should not have any ghosts. If ghosts cannot be removed from the spectrum of quantum states by some mathematically consistent mechanism, then the theory must be discarded (Fig. 7.1).

Fig. 7.1 Negative probability ghosts (Credits, Microsoft Office Collections.)

- Causality violation: This implies a disconnect between cause and effect. An example of causality violation is provided by a hypothetical time machine that would allow you to go back to the past and kill your ancestors before your birth. This is, of course, illogical, and again should not be permitted to occur in a physically acceptable theory. In a theory with extra times, causality violation can occur more easily. So if a way out is not available the theory should be abandoned.

These problems occur even for the first time dimension. A 1T theory is by no means free from these problems automatically. It took half a century to learn how to find the right mechanisms to suppress them in the currently successful 1T theories, including the standard model, general relativity, and string theory.

Avoiding such problems was thought to be notoriously difficult or impossible when more times are present in the formalism. This led most physicists to believe that additional time-like dimensions introduce insurmountable inconsistencies. Furthermore, since it seemed that extra times may not be required to explain what was known, ignoring them seemed to be a preferable and safer choice for most physicists.

However, there were hints of hidden extra time-like dimensions in M-theory, making the issue of extra times not completely irrelevant, thus motivating research in that direction.

The first such hint was noted in my 1995 lectures [3, 4], that a mathematical structure that distinctly exhibited SO(10, 2) symmetry in 10+2 dimensions with two times was evident in the *extended supersymmetry* of M-theory. This is an algebraic construct in 11 dimensions which is of fundamental importance in determining M-theory itself. However, soon after the discovery of 11-dimensional extended supersymmetry by P. Townsend, it quickly became evident to me that this was really a 12-dimensional structure in disguise, giving a clue of an underlying space–time with 10-space and 2-time dimensions.[1] Could this be just a mathematical coincidence or a tip-off for more hidden information about a higher space–time?

An independent clue came soon afterward through F-theory which was proposed by C. Vafa[2] as an extension of M-theory to 12 dimensions. F-theory relied on geometries embedded in 12 dimensions to connect together more facts about string

[1] The clue discussed in [3, 4] is explained in this footnote. The extended superalgebra of M-theory is

$$\{Q_\alpha, Q_\beta\} = P^\mu\left(\Gamma_\mu\right)_{\alpha\beta} + Z^{\mu\nu}\left(\Gamma_{\mu\nu}\right)_{\alpha\beta} + Z^{\mu\nu\lambda\sigma\rho}\left(\Gamma_{\mu\nu\lambda\sigma\rho}\right)_{\alpha\beta}.$$

The supercharge Q_α is the Dirac spinor of SO(10, 1) with 32 *real* components, P^μ is the momentum in 11 dimensions, while the anti-symmetric tensors $Z^{\mu\nu}$, $Z^{\mu\nu\lambda\sigma\rho}$ are called the 2-brane and 5-brane charges, respectively. The 528 charges (P^μ, $Z^{\mu\nu}$, $Z^{\mu\nu\lambda\sigma\rho}$) commute with each other and with the Q_α. This superalgebra is just a rewriting of the following 12-dimensional algebra [3, 4]:

$$\{Q_\alpha, Q_\beta\} = Z^{MN}\left(\Gamma_{MN}\right)_{\alpha\beta} + Z^{M_1M_2M_3M_4M_5M_6}\left(\Gamma_{M_1M_2M_3M_4M_5M_6}\right)_{\alpha\beta},$$

where Q_α is the Weyl spinor of SO(10, 2), which also has 32 *real* components, while Z^{MN}, $Z^{M_1M_2M_3M_4M_5M_6}$ are anti-symmetric tensors in 12 dimensions, with a self-duality condition imposed on $Z^{M_1M_2M_3M_4M_5M_6}$. It is the reality property of the 32 spinor components that dictates that 2 out of the 12 dimensions *must be time-like*. The 32 components of the SO(10, 2) spinor Q_α and the 528 components of the SO(10, 2) tensors Z^{MN}, $Z^{M_1M_2M_3M_4M_5M_6}$ in 12 dimensions exactly reproduce the SO(10, 1) spinor Q_α and the SO(10, 1) tensors (P^μ, $Z^{\mu\nu}$, $Z^{\mu\nu\lambda\sigma\rho}$) when rewritten in an 11-dimensional basis. So the extended superalgebra has a very clear 12-dimensional interpretation. In this way, many key consequences in M-theory that follow from supersymmetry alone can be given a 12-dimensional interpretation with two times. This observation that was made in 1995 begged for an extension of M-theory and is what motivated S-theory in 1996 as a theory in 11-space and 2-time dimensions [5]. S-theory was later instrumental in discovering the fundamental Sp(2, R) gauge symmetry in phase space that gave rise to 2T-physics. 2T-physics, which turned out to stand on its own principles, is applicable to all areas of physics, not only to M-theory. I will return later to the connection between 2T-physics and S-theory or M-theory.

[2] C. Vafa, "F-theory", Nucl. Phys. **B469** (1996) 403, [arXiv:hep-th/9602022].

theory, thus going beyond M-theory. This approach did not take the 12-dimensional space–time seriously and used it mainly as a mathematical convenience. Depending on the application, the 12th dimension of F-theory was sometimes space-like and sometimes time-like.

More algebraic and geometric proposals involving extra times were studied under the titles of S-theory by me [5–9] and U-theory by C. Hull.[3] Like F-theory, these aimed also at connecting facts of string theory, or the little known M-theory, to symmetries or geometric properties of the higher space–time.

While being useful to enlarge the symmetry perspectives of M-theory, none of the above approaches investigated the full-fledged consequences of having extra dimensions of time. They were content with using the extra times mainly as mathematical convenience to discuss symmetries, ignoring the potential disasters of ghosts and causality violation that would be expected to pop-up if the extra times were taken seriously.

So a return to the core difficulties outlined above was needed in order to understand the full significance of the possible extra time dimension.

The origin of the ghost problem is intertwined with special relativity. The concept of relativistically invariant distance, $(\Delta s)^2 = (\Delta\vec{x})^2 - (\Delta t)^2$, explained in footnote 3, introduces a minus sign in the time-like direction, while for all space-like dimensions the corresponding sign is positive. For each additional time-like dimension another minus sign must be introduced in all invariants. Similar minus signs occur in many other quantities as a consequence of consistency with relativity.

The fundamental minus signs enter also in the computation of probability in quantum mechanics, and they invariably lead to the negative probability ghosts in generic relativistic theories *even in 1T-physics*. Adding more space-like dimensions does not cause any such problems, but adding more time-like dimensions introduces more ghosts, thus creating a mountain of lethal problems that seem to crumble the entire structure, at first sight.

There is no chance of winning the battle against such an army of ghosts without sophisticated equipment, such as a gauge symmetry, as in Fig. 7.2.

What is the device that eradicates the negative probability ghosts in 1T-physics? The cure is found only in theories that have the appropriate mix of gauge symmetries. General relativity, the standard model, and string theory all have just the required amount of gauge symmetry to eliminate the ghosts for 1T while still being consistent with relativity. However, generic theories in 1T-physics that do not have the appropriate gauge symmetry all contain ghosts and are therefore discarded.

So gauge symmetry in 1T-physics plays a very important dual role. On the one hand it banishes ghosts and on the other hand it predicts the presence of a force as well as all of the details of the interactions produced by the force. Indeed one can say that the force exists in order to satisfy the gauge symmetry. The strong, weak,

[3] C.M. Hull, "Timelike T duality, de Sitter space, large N gauge theories and topological field theory", JHEP **9807** (1998) 021, [arXiv:hep-th/9806146].

Fig. 7.2 Sp(2, *R*) gauge symmetry destroys negative probability 2T ghosts (Credits, Bora Orcal.)

electromagnetic, and gravitational forces are all explained in this way, as originating from a corresponding, carefully chosen, gauge symmetry.

1T-physics provided the lessons for how to embark on a search for a consistent mechanism for banishing the ghosts due to more time dimensions. The cure had to be a gauge symmetry, but which one? It could not be either the familiar general coordinate transformations or the usual Yang–Mills-type transformations of field degrees of freedom. Those are just sufficient to eliminate the ghosts of a single time dimension, they cannot go beyond. So, one had to look into the clues given by M-theory or its extensions to search for a new symmetry.

Some initial steps which pointed in new symmetry directions hinted by S-theory were taken by me in collaboration with C. Kounnas during 1996–1997 [11, 12] and with my student C. Deliduman [13, 14]. That work introduced new gauge symmetry concepts to eliminate ghosts for a system of two particles in a space–time with one extra time dimension. While enlightening, the approach with two or more particles was fundamentally unsuitable to describe a *single* particle in a space–time with extra time dimensions without running into the usual problems.

Finally, while attending a conference at CERN at the end of 1997, I discovered the appropriate new gauge symmetry based on symplectic transformations called Sp(2, *R*). This Sp(2, *R*), which will be explained in detail in the next section, is the symmetry that transforms all of phase space degrees of freedom (position, time, momentum, energy) into each other and makes them indistinguishable at every instant of motion. After discussions with my student C. Deliduman and postdoc O. Andreev, we presented the concept in a 1998 paper in a greatly simplified context of a single spinless particle [15].

The new symmetry is possible only if space–time has exactly two times. Neither one time nor three times are acceptable on mathematical or physical grounds. It is fundamentally a new type of gauge symmetry acting in phase space. That is, it mixes the generalized coordinates of space–time–momentum–energy, not only space–time, and puts them all on an equal basis. This gauge symmetry has the right ingredients to fit the earlier hints from S-theory and M-theory, a clear encouragement for taking it seriously in the context of M-theory.

The Sp(2, *R*) gauge symmetry principle turned out to be fundamental in its own right and was later generalized to more complicated systems [48–51], including particles with spin [17], particles with supersymmetry [19], particles interacting with a background of various fields [23, 24], and finally it was also implemented in field theory [25] which is the proper language [37] for the standard model [38] and general relativity [43].

Over time, it became apparent that Sp(2, *R*) gauge symmetry is a broad principle applicable generally to all physics, not only to the mysterious fundamental physics that M-theory and other approaches aim to understand. It is relevant for formulating the fundamental laws of physics as well as the effective laws that describe everyday physics, and it yields more physics information that the usual approach in 1T-physics is unable to capture or predict. Consequently, the new gauge symmetry offers a deeper perspective to understand our universe at all scales, from the very small to the very large.

7.2 Indistinguishable Position and Momentum and Sp(2, *R*)

Perspectives of observers play an important role in how nature appears to them. The examples in Figs. 1.2, 1.3, 3.5, 3.6 and 4.1 illustrate this point. On the other hand, symmetries determine the invariant quantities that are independent of the perspectives, so that observers connected by the symmetry transformations always measure the same identical invariant quantities.

An elementary discussion of symmetry and perspective was presented in Section 7.4. Some examples of relativistic invariants that we mentioned before, for inertial observers only, include the unchanging velocity of light c and the relativistically invariant distance (footnotes 1, 3 in Section 3.1).

Past successes of theoretical physics relied heavily on discovering and then taking advantage of the symmetries that govern nature. The symmetries restrict the form of the equations that describe the fundamental laws, producing mathematical expressions that are unique in some cases.

In particular, as discussed before, the notion of gauge symmetry that every local observer can verify at his own position in space–time, independently from observers elsewhere, turned out to be extremely powerful, so powerful that the very existence of the fundamental forces as well as all details of the interactions mediated by them emerged from the gauge symmetry.

Now we ask, have physicists found all the symmetries of nature and accounted for all possible observers? Surprisingly, after so many years of research in symmetries, the answer is no! There is more symmetry, and more fundamental laws, to be discovered in the future. That this is the case became evident with the discovery of the Sp(2, R) symmetry in phase space, which in turn led to 2T-physics. This opened up new perspectives of observers that had not been dreamed of before, and this avenue of research is only at the beginning.

What are the new perspectives? When we discussed perspectives earlier, we meant perspectives of observers in various configurations of space–time as depicted in Figs. 3.1, 3.2, 3.3, 6.3 and 6.4. Space–time is specified with coordinates $X^M = (t, x, y, z, t', x', \ldots)$, including the possible extra time t' and extra space x' dimensions required by 2T-physics. The gauge symmetries in the successful theories of the past considered local transformations of degrees of freedom in position space X^M. A super simplified example of local transformation is rotating the tiles individually in the example of Fig. 4.1, but now imagine not only local in space but also in time, including the extra dimensions represented by X^M as well.

However, space–time X^M is only half of the variables that are needed to describe the state of motion of any system or of any observer. The other half are the momentum and energy variables $P^M = (E, P_x, P_y, P_z, E', P_{x'}, \ldots)$, including a possible extra energy E' or momentum $P_{x'}$ associated with the extra dimensions of 2T-physics. The ellipsis "... " stands for additional possible extra dimensions of space–time both for position and momentum spaces.

Extending the notions of relativity to include the extra dimensions, all the X^M represent *relativistic position space*, while all the P^M represent *relativistic momentum space*. The full set X^M, P^M is called *relativistic phase space*. All of phase space is needed to describe motion at any instant.

This larger set of variables in phase space offers additional perspectives from which to view a physical system. The symmetries in the successful theories of the past are unable to connect the new perspectives just mentioned to previously considered perspectives. This is because *the known gauge symmetries do not mix relativistic position space X^M with relativistic momentum space P^M*.

Natural questions arise: Is it possible to consider more general transformations in phase space X^M, P^M to build new gauge symmetries? Would those be consistent with the known fundamental laws of nature? What new information would they reveal?

The starting point is a re-examination of the classical or quantum laws of motion at the fundamental level. One can then note that there is a point of departure that is completely symmetric under the interchange of covariant position X^M with contravariant momentum P_M.

Indeed these variables appear at the same level of importance in the setup of classical mechanics in the form of *Poisson brackets* or in the setup of quantum mechanics in terms of *quantum commutators*. Furthermore, X^M and P_M are also at the same level of importance in specifying boundary conditions or in reporting the outcome of *any measurement*.

As long as we focus on these general aspects of motion, rather than properties of specific systems, we can identify the following symmetry under the interchange of relativistic position and momentum:

$$\text{Symmetry:}\, X^M \to P_M \text{ and } P_M \to -X^M.$$

That is, *nothing in the laws of classical or quantum mechanics changes* if relativistic space–time is replaced by momentum–energy and vice versa.

Of course, a particular dynamical system (i.e., its Hamiltonian) turns into another one under this transformation, but *we will attribute the dissimilarity of the systems as being due to the different perspectives of diverse observers*, rather than regarding them as fundamentally different dynamical systems. This introduces the new point of view that goes beyond the common formulation of physics.

The example of position–momentum interchange symmetry given above now provides a glimpse of the types of new perspectives that we wish to relate by the new symmetry transformations. This is an example of a *canonical transformation* in phase space. Actually, it is only a small part of the possible symmetries based on canonical transformations hinted by the action principle, as described in footnote (4). However, this symmetry is not enough. The *drawback* in this special case, as well as its generalizations to all other familiar canonical transformations, is that these are only *global* transformations. To solve the ghost problem, what is needed is a symmetry under *local* gauge transformations.

It turned out that a small subset of all canonical transformations, called Sp(2, R), could be promoted into a gauge symmetry that is local at *each instant* of the worldline of a moving particle, no matter how complicated its motion is. The emphasis here is that the new symmetry is *local*,[4] and therefore has the power of curing the ghost and causality problems of extra time-like dimensions.

The Sp(2, R) local symmetry is the only subset of canonical transformations that has worked consistently to eliminate all the ghosts due to two time-like dimensions. Canonical transformations larger than Sp(2, R) could not be turned to gauge symmetries: either they were too strong and collapsed the system to nothing or they required too many time-like dimensions without being sufficient to remove all ghosts. Hence the phase space gauge symmetry approach works only with Sp(2, R) and only with two times, no less and no more [47].

The new gauge symmetry principle, described graphically in Fig. 7.3 and mathematically in footnotes 5, 6, is broader than previous gauge principles because it transforms observers in relativistic phase space (X^M, P_M), not just in relativistic

---

[4] The action principle and the global canonical transformations that we wish to extend to local symmetries are best illustrated for a point particle without spin. The action that determines its motion is $S = \int d\tau \left(\frac{dX^M}{d\tau} P_M - H(X,P) \right)$. A given Hamiltonian $H(X, P)$ specifies a particular system. If we do not focus on a specific system we can ignore H to analyze the general properties of phase space. The first term of the action $S_0 = \int d\tau \frac{dX^M}{d\tau} P_M$ dictates the properties of position versus momentum, including their roles in Poisson brackets or quantum commutators. This first term has the symmetries of all canonical transformations. This is a huge *global* symmetry that includes the special cases mentioned in the text. It is a global rather than a local symmetry on the worldline because its generators do not depend explicitly on the worldline parameter τ which tracks the worldline. Parts of this global symmetry can be promoted to gauge symmetries that are local on the worldline as described in footnotes 5, 6.

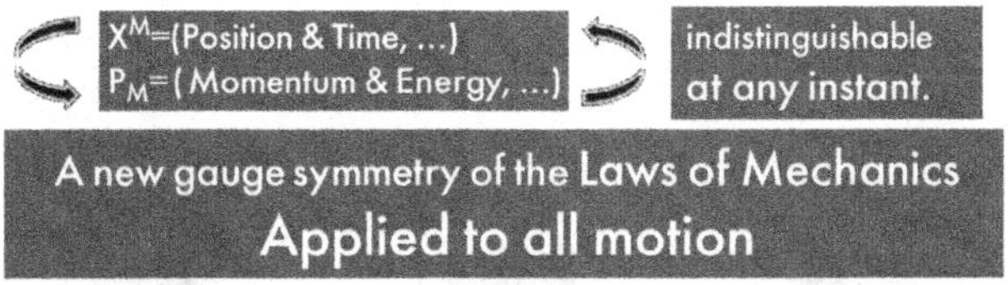

Fig. 7.3 Relativistic momentum–position *gauge* symmetry at each instant

position space X^M. This type of local symmetry was not dreamed of before the advent of 2T-physics.

Recall that the worldline represents the history of the motion of the particle, as in Fig. 6.4. That figure described the history of events from the perspective of position space, thus favoring position over momentum in the description of motion. The Sp(2, R) symmetry makes position space X^M and momentum space P_M indistinguishable from each other *at every instant along the worldline*. Then observers with perspectives all over phase space, not only in position space, will determine the same Sp(2, R) invariants.

The Sp(2, R) gauge symmetry turned out to be the right prescription to cure both the ghost problem and the causality problem for all possible motions of a particle without spin. The local symmetry is generalized for more complicated systems, such as spinning particles, supersymmetric particles, or particles moving in background fields, and eventually field theory.

It is remarkable that the Sp(2, R) gauge symmetry not only agrees with all known facts of physics but also goes further to uncover new correlations between perspectives in phase space that were not suspected before. The previously hidden correlations in 1T-physics become evident in 2T-physics because the Sp(2, R) gauge symmetry enforces a higher perspective in a space–time with one extra space and one extra time dimensions.

7.3 How Does It Work?

The symmetry has multiple roles. One of them is to make 2T-physics compatible with 1T-physics while unifying various systems in 1T-physics. The shadows allegory described in Section 5.2 and Figs. 5.2, 7.4 and 7.7 is very useful to grasp the concepts that follow, so the reader is urged to review the shadows allegory in Section 5.2 at this point.

In this section we discuss specifically the example of a spinless point particle in 2T-physics, but the concepts are general and are known to apply all the way to 2T-field theory that describes the 2T versions of the standard model and general relativity. The crucial concepts are summarized in Figs. 7.4, 7.5, 7.6 and 7.7.

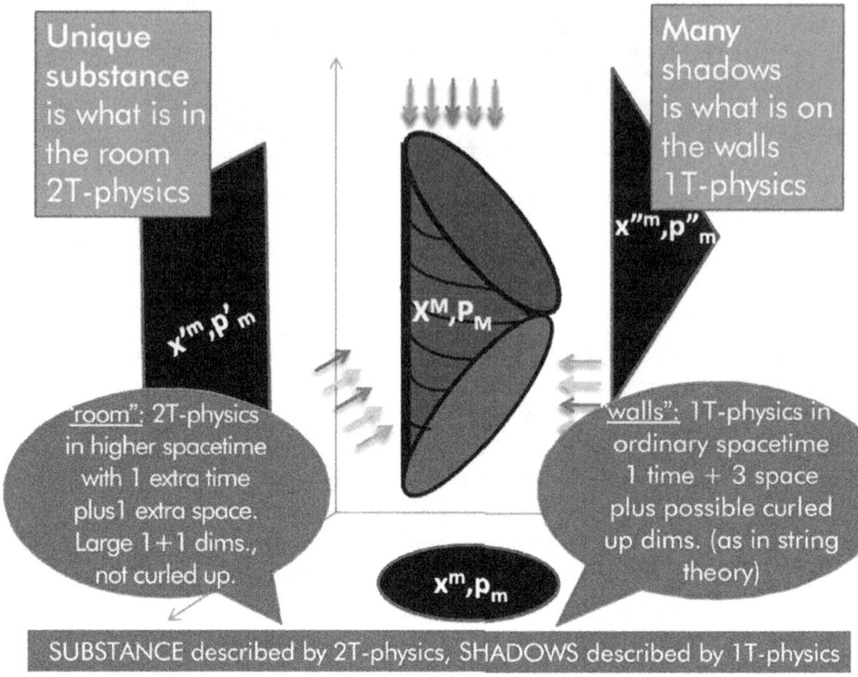

Fig. 7.4 2T-physics in $d + 2$ dimensions produces many 1T-physics "shadows" in $(d - 1) + 1$ dimensions. 2T-physics has one extra space and one extra time dimensions. The allegory of a substance in a room and its "shadows" on surrounding walls is helpful to describe the relation between 1T-physics (shadows) and 2T-physics (substance)

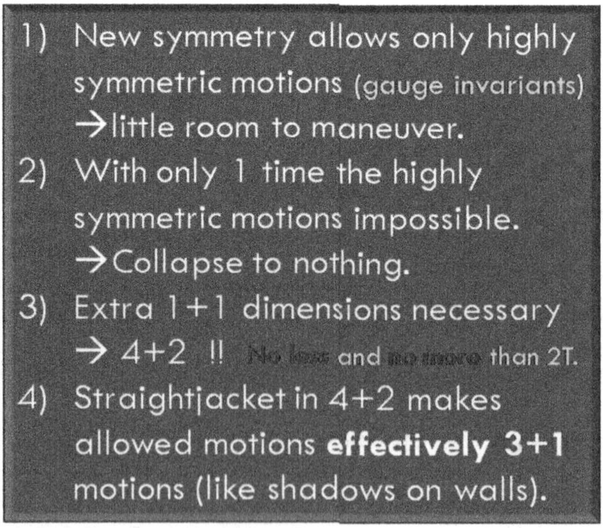

Fig. 7.5 Sp(2, R) symmetric motions possible only in 4 + 2 dims

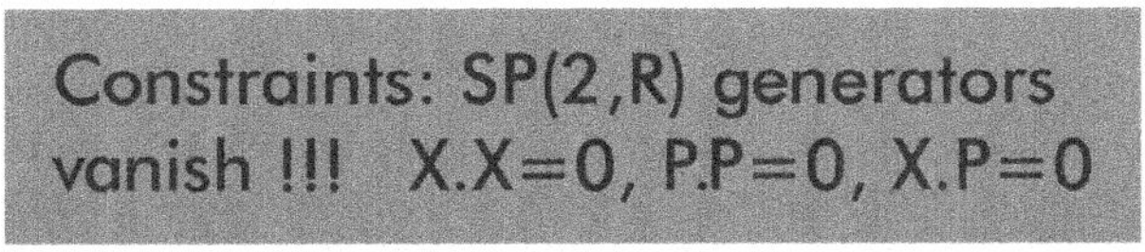

Fig. 7.6 Vanishing Sp(2, R) generators in flat space–time. These are special forms of the more general equations $Q_{ij}(X, P) = 0$ mentioned in footnote 5

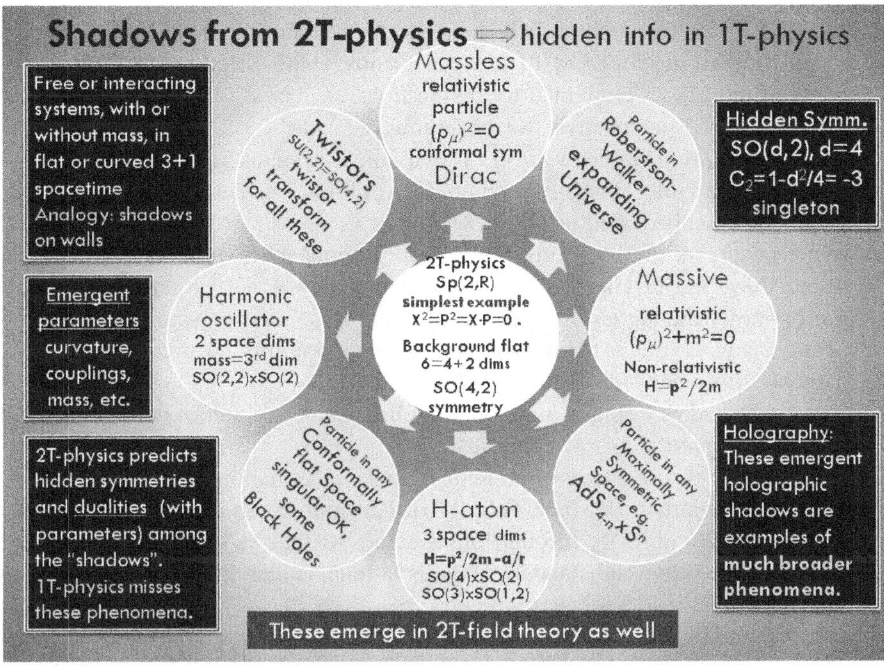

Fig. 7.7 Shadows in 3 + 1 capture different perspectives of 4 + 2. This is analogous to many takes of a movie scene from different perspectives. The dualities implied by this figure are analogous to the dualities in M-theory in Fig. 10.1

7.3.1 Informal Discussion

In analogy to Fig. 7.4, "substance" is the phase space of a particle in 4-space and 2-time dimensions (X^M, P_M), while the solutions of the Sp(2, R) Sp(2, R) constraints in Figs. 7.5 and 7.6 will be the 3 + 1-dimensional "shadows," as given in Fig. 7.7.

The summary in Fig. 7.5 gives a brief outline of how 2T-physics and 1T-physics become compatible: the Sp(2, R) symmetry puts a "straightjacket" on the system in 4 + 2 dimensions that restricts possible motions only to an allowable symmetric subset of phase space. The fully symmetric motions demanded by these constraints cannot exist in a space–time with only 1 time. So to have non-trivial content and for the symmetry to work to remove ghosts, the starting point has to be a space–time with two times, no less and no more. The allowed symmetric motions in

4 + 2 dimensions are so restricted that they can effectively be interpreted as being equivalent to motions in 3 + 1 dimensions.

Different 3 + 1 shadows of the same 4 + 2 phase space emerge just like different shadows of the same object in a 3-dimensional room get created by shining light on it from different perspectives, as in Figs. 7.4 and 7.7.

This effective reduction to 3 + 1 dimensions is not trivial, because the relation between "substance" and "shadow" is not one-to-one. Rather, many 3 + 1-dimensional shadow histories (as in Fig. 3.3) emerge for the same space–time history in 4 + 2 dimensions. The different 3 + 1 shadows represent different dynamical systems described by 1T-physics formulas with different Hamiltonians or equations of motion, as explained following Fig. 7.7. These *multiple shadows* provide different 3 + 1 perspectives for observing the same 4 + 2-dimensional *unique system*. In this way a unification of 1T dynamical systems is achieved: shadows that belong to the same substance are naturally unified. The details are predicted by 2T-physics and can be tested in 1T-physics.

We are observers in 3 + 1 dimensions, in the same space–time as the shadows. We do not have the privilege of observing directly in 4 + 2 dimensions and see straightforwardly the higher dimensional system. We can only examine the properties of the shadows in 3 + 1 dimensions and have no ability to directly investigate the original "substance" in 4 + 2 dimensions.

Since each shadow contains only one timeline, causality within the dynamics of the shadow is as natural as in 1T-physics.

These distinct 3 + 1 shadows contain leftover information from 4 + 2 dimensions for observers that are stuck in 3 + 1 dimensions. Such observers experience 4 + 2 dimensions indirectly through many relations that exist among the various "shadows" of the same "substance." Those relations, called dualities, are predicted precisely by the built-in Sp(2, R) symmetry.

Sp(2, R) provides the tool to compensate for our underprivileged position as observers on the "wall." The information about the higher space–time is not lost, but it remains embedded in the properties of the shadows. Sp(2, R) predicts the precise relations among the shadows, thus guiding observers in 3 + 1 dimensions for what to look for and how their observations should be related to the perspectives in 4 + 2 dimensions.

The formalism of 2T-physics makes precise predictions for 3 + 1 observers of what these properties ought to be. So, they can be brought out by recognizing that dynamical systems in 3 + 1 dimensions have previously unsuspected hidden symmetries and relations among themselves that reveal their properties inherited from the higher space–time in 4 + 2 dimensions.

For more specifics on these notions see the discussion following Fig. 7.7.

7.3.2 *Formal Discussion*

Having described informally how the symmetry works in Layman's terms, we now give the outline of a more formal explanation. For a complete mathematical explanation refer to my papers.

The mathematical formulation of the Sp(2, R) gauge symmetry and the emergence of the constraints in Fig. 7.6 is described in footnote 5. A universal essential property of any *gauge symmetry* is that it severely restricts the allowed physical configurations that the system can exist in. The Sp(2, R) symmetric system[5] is subject to *new restrictions that supplement the laws of motion in classical or quantum mechanics*. Specifically, in the simplest example of *flat space–time*,[6] Sp(2, R) gauge invariance imposes the constraints $X\cdot X=0$, $P\cdot P=0$, and $X\cdot P=0$ shown in Fig. 7.6. These are above and beyond any equations of motion of a specific dynamical system. These additional equations represent the "straightjacket" alluded to in item 1 in Fig. 7.5. Their significance is to impose indistinguishable position–momentum at every instant of motion in flat space–time. That is, the solution space of these constraints corresponds to the *Sp(2, R) gauge-invariant subspace of the full phase space*. In the absence of the Sp(2, R) symmetry all phase space configurations X^M, P_M in 4 + 2 dimensions could occur but, with the symmetry, only the Sp(2, R) gauge-invariant configurations of X^M, P_M that satisfy the constraints are admissible.

---

[5] The laws of motion that include some unique formulas are dictated by the new Sp(2, R) gauge-invariant action principle. The gauge principle requires that every derivative must be replaced by a gauge covariant derivative. In particular, the τ derivative $\frac{dX^M(\tau)}{d\tau}$ that appears in the action in footnote 4 must now be replaced by a covariant derivative. Consequently the first term of the action now gets replaced by $S=\int d\tau\left\{\frac{dX^M(\tau)}{d\tau}P_M-\frac{1}{2}A^{ij}(\tau)\,Q_{ij}(X,P)\right\}$ where $Q_{\mathrm{ij}}(X,P)$ are the generators of the Sp(2, R)transformations and $A^{ij}(\tau)\,\left(=A^{ji}(\tau)\right)$, a symmetric 2×2 matrix, with i, j=1, 2) are the gauge potentials. This modified action now incorporates the consequences of the gauge symmetry and determines the rules of mechanics accordingly. The form of the $Q_{ij}(X,P)$ as a function of (X, P) depends on the forces that are being applied on the particle in d + 2 dimensions. All possible $Q_{ij}(X,P)$ permitted by the Sp(2, R) symmetry are known generally for all physical circumstances. The simplest form occurs when the particle moves in *flat space–time* X^M with no forces other than the symmetry constraints. In this case $Q_{11}=X\cdot X$, $Q_{12}=Q_{21}=X\cdot P$, and $Q_{22}=P\cdot P$ as in Fig. 7.6. In this special case the Sp(2, R) transformations between two perspectives are linear as described in footnote 6. Minimizing the action S with respect to the gauge potential A^{ij} gives the subset of equations of motion which correspond to the constraints $Q_{ij}(X,P)=0$ as in Fig. 7.6.

[6] In flat space the Sp(2, R) transformations take the special linear form that can be written as a 2×2 real matrix of determinant 1, as follows:

$$\begin{pmatrix}\tilde{X}^M(\tau)\\ \tilde{P}^M(\tau)\end{pmatrix}=\begin{pmatrix}a(\tau) & b(\tau)\\ c(\tau) & \frac{1+b(\tau)c(\tau)}{a(\tau)}\end{pmatrix}\begin{pmatrix}X^M(\tau)\\ P^M(\tau)\end{pmatrix}.$$

The matrix above contains $a(\tau), b(\tau), c(\tau)$ as its three transformation parameters that are local on the worldline parametrized by τ. Here the doublet $X^M(\tau), P^M(\tau)$ represents the changing phase space along the worldline from the perspective of one observer, while the doublet $\tilde{X}^M(\tau), \tilde{P}^M(\tau)$ is the perspective of another observer. The two perspectives are related by the symmetry transformation above, such that $\tilde{X}^M(\tau)=a(\tau)X^M(\tau)+b(\tau)P^M(\tau)$ and $\tilde{P}^M(\tau)=c(\tau)X^M(\tau)+\frac{1+b(\tau)c(\tau)}{a(\tau)}P^M(\tau)$. The gauge fields $A^{ij}(\tau)$ must also be transformed as triplets to verify the symmetry of the action in footnote 5. Each observer at any instant τ can choose the values of $a(\tau), b(\tau), c(\tau)$ arbitrarily; so the transformation of an observer to a new perspective at instant τ is independent than the transformation of another observer at a different instant τ'. This freedom of observers is the property of a gauge symmetry that is *local on the worldline*. When the particle is subject to various forces, the Sp(2, R) transformations from (X^M, P^M) to $\left(\tilde{X}^M, \tilde{P}^M\right)$ become more complicated, but they are well understood for all possible cases.

This subspace has the same number of phase space degrees of freedom as if the motion takes place in 1 less time and 1 less space dimensions.

The constraints in Fig. 7.6 provide an example of the form of the three generators of Sp(2, R) symmetry $Q_{ij}(X, P)$, with $i, j = 1, 2$. This form of $Q_{ij}(X, P)$ is appropriate for flat space–time in $d + 2$ dimensions. When the particle interacts with background fields, such as gravity, electromagnetism, or more complicated cases, the three constraints in Fig. 7.6 are replaced by three more general functions of phase space $Q_{11}(X, P)$, $Q_{22}(X, P)$, $Q_{12}(X,P)$ (with $Q_{21} = Q_{12}$) that form the Sp(2, R) Lie algebra under Poisson brackets as described in footnotes 5, 6. The generalized form of the $Q_{ij}(X, P)$ includes motion in curved space–time in $d + 2$ dimensions. The equations of motion require the three generalized gauge generators to vanish $Q_{ij}(X, P)=0$, as in the flat space example of Fig. 7.6. When the symmetry generators vanish *the corresponding phase space that solves these equations is gauge invariant*. So the equations in Fig. 7.6 and their generalizations amount to demanding an Sp(2, R) gauge-invariant physical sector.

These equations are so constraining that only special configurations of phase space X^M, P_M can satisfy them. In a space–time with only one time these highly symmetric gauge-invariant motions are impossible, so that Sp(2, R)-invariant 1T dynamics cannot exist as summarized in Fig. 7.5 and mathematically explained in footnote 7.

We see that the requirement of vanishing Sp(2, R) symmetry generators $Q_{ij}(X, P) = 0$, which implies that only Sp(2, R) gauge invariants are physical, is the underlying reason for having two time dimensions. There is no non-trivial dynamics unless there are two times.[7] Evidently, the extra time is not introduced "by hand, " the symmetry demands its existence.

---

[7] To understand this, one must be aware that there is a space–time metric in the definition of the dot products in Fig. 7.6, e.g., $X \cdot X \equiv X^M X^N \eta_{MN}$, where η_{MN} is the flat space–time metric. So the meaning of $X{\cdot}X$ is $X{\cdot}X = -t^2+x^2+y^2+z^2-(t')^2+(x')^2+\cdots$, and similarly $P{\cdot}P = -E^2+p_x^2+p_y^2+p_z^2-(E')^2+(p_{x'})^2+\cdots$ and $X \cdot P = -tE+xp_x+yp_y+zp_z-t'E'+x'p_{x'}+\cdots$. Note the minus signs in the two time-like directions t, t' or E, E'. In the case of 1T-physics the extra terms with t',E',x',$p_{x'}$ are absent. The equations $X{\cdot}X{=}P{\cdot}\ P{=}X{\cdot}\ P{=}0$ in 2T-physics have solutions only if there are two times; with fewer times these conditions collapse the system to triviality. To see why, consider at first 0-time dimensions, and ask if there are solutions to equations such as $x^2+y^2+z^2+(x')^2=0$ when there are no minus signs. Then the only solution is trivial $x{=}y{=}z = x' = 0$, that is vanishing vectors $X^M{=}P^M{=}0$ which has no physical content. Next consider 1-time, and find that the only solution is that $X^M{\sim}\ P^M$ must be parallel light-like vectors *proportional* to each other. In that case the angular momentum vanishes $L^{MN}{=}X^M\ P^N{-}X^N P^M{\Rightarrow}0$, and hence this is a trivial solution that does not represent even free motion in 1T-physics. In 2T-physics, the presence of the extra time coordinate permits an infinite number of non-trivial solutions. These solutions represent the Sp(2, R) gauge-invariant sector of phase space. These solutions are parametrized by phase spaces in 1 less time and 1 less space dimensions x^μ, p^μ, which describe various types of interacting particle dynamics as in the examples of Fig. 7.7. The lower dimensional phase spaces associated with these distinct 1T dynamical systems parametrize the "shadows" of the higher dimensional phase space X^M, P_M. All the gauge-invariant information, which is guaranteed by the vanishing generators $Q_{ij}{=}0$, is captured holographically by each shadow.

Once a space–time with two times is admitted, these constraints yield a large set of solutions. As shown in the examples of Fig. 7.7, each solution turns out to have an interpretation in terms of 1T-physics in a phase space with one less time and one less space dimensions.

But isn't a theory with two times problematic because of the ghost and causality problems discussed before? Yes, these would have been problems in the absence of a gauge symmetry. However, the Sp(2, *R*) gauge symmetry is the right medicine to cure those problems. Indeed, Sp(2, *R*) has precisely the correct amount of gauge symmetry to banish all the negative probability ghosts due to both the first and second time dimensions.

We have already indicated that causality works just like 1T-physics because each shadow contains only one timeline, hence 1T observers see no causality violation.

The solutions of the constraints, namely the shadows, are ghost free, satisfy causality, and furthermore each one of them captures holographically all of the gauge-invariant information of the higher dimensional phase space.

In all known cases of physically successful gauge symmetries, the symmetry plays a dual role. On the one hand it demands the existence of a certain structure of the physics equations, making them unique. On the other hand it eliminates ghosts. Likewise, Sp(2, *R*) demands an extra space and time dimensions, with a specific set of equations, while also being just sufficient symmetry to eliminate ghost and causality problems.

These magical properties of Sp(2, *R*) cannot take care of the problems if there are more than two times. With more times there are more ghosts, but insufficient symmetry to eliminate them. Therefore in a theory with Sp(2, *R*) gauge symmetry there must be *precisely two times, no less and no more.*

Moreover, a larger gauge symmetry that would permit more than 2 times *without ghosts* does not seem to exist in the context of particle dynamics that generalize the approach in footnotes 5, 6. The balance between the number of ghosts, amount of symmetry, and also the existence of non-trivial solutions of the constraints as explained in footnote 7 have so far been impossible to satisfy in many attempts, except for the case of Sp(2, *R*). So, Sp(2, *R*) appears to have a special status which makes it more appealing [47].

7.4 Examples of Shadows and Hidden Information

The previous section provided a general overview of how 1T-physics in 3 + 1 space–time emerges in the form of shadows that get unified by 2T-physics in 4 + 2 space–time.

It is very instructive to consider an example of a 2T-physics system and how it relates to certain 1T-physics systems. The simplest concrete example is summarized in Fig. 7.7. It is the spin less particle in flat space–time in 4 + 2 dimensions subject to the Sp(2, *R*)) gauge symmetry constraints of Fig. 7.6, but otherwise a free system in the absence of any forces, as formulated in footnotes 5, 6, 7.

The central blob in Fig. 7.7 represents the "substance" in 4 + 2 dimensions subject to the Sp(2, R) gauge symmetry. Because this system is in flat 4 + 2 space–time, it also has the global symmetry SO(4, 2) associated with observers connected to each other by special relativity in 4 + 2 dimensions. This is the Lorentz symmetry in 4 + 2 dimensions that is evident in the dot products of the constraints $X \cdot X = P \cdot P = X \cdot P = 0$ recorded in the central blob of Fig. 7.7.

The surrounding blobs represent examples of 1T physics systems that emerge as 3 + 1 "shadows" of the same "substance" in 4 + 2 space–time. These are solutions to the constraints $X \cdot X = P \cdot P = X \cdot P = 0$. The interpretation of the emergent dynamical system for each shadow in 1T-physics is recorded in the corresponding blob.

Although the system in flat space–time in 4 + 2 dimensions is very simple, with no dynamics other than the constraints, the solutions in the form of shadows live in emergent space–times in 3 + 1 dimensions with a rich variety of dynamics, hidden symmetries, and dualities that materialize from gauge fixing the 2T theory in various ways. "Gauge fixing" is a mathematical procedure used for solving the equations and arriving at the lower dimensional space–time. The gauge fixing provides many ways of embedding phase space in 3 + 1 dimensions into phase space in 4 + 2 dimensions. Although a 2T-physicist would consider this gauge fixing to be just a convenience, it has deep physical implications to a 1T-physicist because a choice of time and of Hamiltonian that governs the dynamics of the shadow emerges in the gauge fixing process. While a 1T-physicist considers the different Hamiltonians as distinct systems, from the point of view of the 4 + 2 formalism they are merely different looking, but *gauge-invariant phase space perspectives* for the same 4 + 2 system.

Some examples of shadows of the flat 4 + 2 system that have been worked out in detail over the past 10 years are the following cases; some of which are shown also in Fig. 7.7. These examples are not exhaustive of all possible solutions of the simple form of the constraints[8] in Fig. 7.6. The mathematical expressions for each shadow described below in words are summarized with formulas in Tables I, II, and III of [41].

- The massless relativistic particle in 3 + 1 flat Minkowski space. This shadow describes a particle without mass, such as photon, moving at the speed of light without any forces being applied on it.
- The massive relativistic particle in 3 + 1 flat Minkowski space. This shadow describes a particle with mass, such as an electron, moving without any forces being applied on it, but obeying the rules of relativity as in Fig. 3.4. Evidently, it is a different shadow since the 1T observer can measure the mass.

[8] The more general forms of the constraints $Q_{ij}(X, P) = 0$, for example, in the presence of background fields for gravity or electromagnetism [23, 24, 43], have been little analyzed. These more general cases would yield their own more interesting shadows.

- The non-relativistic free massive particle in 3 space dimensions. This is also the free massive particle, but its dynamics obeys Newton's laws of motion, so that its energy–momentum relation is $E = \vec{p}^2/2m$ and the velocity is $\vec{v} = \vec{p}/m$ which are different than the relativistic relations in Fig. 3.4. Again this shadow is distinguishable from the above ones by 1T observers.
- The non-relativistic hydrogen atom (i.e., α/r potential) in 3 space dimensions. This shadow describes the motion of a particle with mass m, moving around a center under the influence of an attractive radially directed force $F = \alpha/r^2$ that decreases like one over the distance-squared. Like the mass m, the strength of the force α emerges also as a parameter that defines the shadow.
- The harmonic oscillator in 2 space dimensions, with its mass interpreted as a third dimension. This shadow describes a massive particle performing oscillatory motion on a 2-dimensional surface under the influence of an attractive radially directed force $F = \alpha r$ that increases linearly with distance r.
- The particle moving on the Robertson–Walker cosmological space–time. This shadow describes a massless particle, such as light, moving under the influence of an expanding universe in 3 + 1 dimensions. There is an effective force applied on the particle due to the expansion of the fabric of space and time. To get a feeling about this motion, consider the surface of an expanding balloon and watch the motion of two points marked on the balloon with a pen.
- The particle moving on a space–time generated by a cosmological constant. This is also the motion in a curved space–time, the one that would result if at every point in space–time there is a constant energy density and pressure created by the cosmological constant.
- The particle moving on the curved space AdS_4 or on dS_4. These shadows describe the motion of a massless particle in a special curved space–time in 3 + 1 dimensions called deSitter (dS_4) or anti-deSitter (AdS_4). The particle moves on curved trajectories called the geodesics on these spaces.
- The particle moving on the curved space $AdS_3 \times S^1$ or on $R \times S^3$. These are also curved spaces. $AdS_3 \times S^1$ is anti-deSitter in 2+1 dimensions (AdS_3) plus a circle (S^1) in 1 space dimension. $R \times S^3$ is the infinite line for the time coordinate while S^3 is a "sphere." The particle moves on the geodesics of these spaces.
- The particle moving on the curved space $AdS_2 \times S^2$. Similar to the above, where AdS_2 is anti-deSitter in 1+1 dimensions, while S^2 is the 2-dimensional surface of sphere.
- The particle moving on any maximally symmetric space of positive or negative curvature. This is a class of curved space–times.
- The particle moving on any conformally flat space–time, including singularities. The shadow describes the motion of the particle on the geodesics of a conformally flat space that may or may not include singularities.
- A related family of other particle systems, including some black hole backgrounds.

It should be noted that among the shadows one finds free or interacting systems describing particles, with or without mass, and moving in either flat or curved space–times in 3 + 1 dimensions. Therefore some parameters which were not initially in 4 + 2 dimensions emerged in the solutions for the shadows. These emergent parameters include mass, strength of interaction, curvature.

Could the higher dimensional mechanisms of 2T-physics lead to a new understanding of the origin of mass, various interactions, or curved spaces of cosmological significance? It should be noted that the standard Robertson–Walker expanding universe in 3 + 1 dimensions, which is of fundamental importance in cosmology, is included among the solutions of this very simple system that comes as a shadow from 4 + 2 dimensions. Such questions are currently being investigated.

Figure 7.7 is reminiscent of a similar figure that summarizes dualities in M-theory as in Fig. 10.1. In fact there are further points to be mentioned as being analogous to M-theory. These include the emergent parameters such as mass, curvature which are reminiscent of similar parameters called "moduli" in M-theory. Additional features include holography and dualities that are explained below.

A general property of each shadow is that it is *holographic*. This means that it contains all of the information of the original substance in 4 + 2 dimensions even though it itself is in 3 + 1 dimensions. So, in principle, a 3 + 1 observer could gather all that information if he/she can figure out how to do it. 2T-physics provides the missing information for how to proceed.

Another general property is *duality* which is related to the holographic property. Two systems are considered duals of each other if by studying one of them we are able to completely figure out the other, at least in principle. Since each shadow is holographic it contains all the information whether it is in the form of one shadow or another. So duality follows from the holographic property. Duality implies unification of different systems into one.

By the very construction of 2T-physics, there must be specific hidden relations among the various 3 + 1 shadows since they come from the same 4 + 2 substance, while each shadow is holographic. The differences among shadows from the point of view of 1T-physics are attributed to the different 3 + 1 perspectives in phase space. Therefore there has to be duality transformations between these different 1T-physics dynamical systems. In the 1T formulation there is no hint that some systems are dual to one another, but 2T-physics makes predictions for 1T-physics through the shadows and furthermore specifies that the duality transformations are Sp(2, R)) gauge symmetry gauge transformations that relate the different 1T dynamical systems. So the duality transformations among all the systems discussed above are precisely given by the linear Sp(2, R) gauge transformations discussed in footnote 6. Special choices of the parameters $a(\tau)$, $b(\tau)$, $c(\tau)$ relate the different shadows to each other.

The duality property of the examples in Fig. 7.7 has been verified in this way by using directly 1T-physics methods, thus showing that 2T-physics is indeed a unifying structure that stands above 1T-physics. For an example of a duality transformation that relates the massless and massive relativistic particles, see Eq. (58) in

[51]. 1T observers need only follow these predictions of 2T-physics and verify that those dualities are indeed true.

An example of the type of holographic information is the global SO(4, 2) symmetry[9] of the substance in the central blob in Fig. 7.7. What this SO(4, 2) symmetry says is that *observers in 4 + 2 dimensions* see no distinction in the laws of 2T-physics with regard to orientations in 4 + 2 flat space–time. As in special relativity, all directions in space–time are equivalent to observers that may reorient themselves in any of the 4 space directions or boost themselves to moving frames at any constant velocities while measuring time in either or both time directions.

An observer in 3 + 1 dimensions must be able to uncover the effects of this 6-dimensional symmetry since, by holography, this must be a hidden symmetry of each shadow. So, 2T-physics predicts that each shadow in Fig. 7.7 must have a hidden SO(4, 2) symmetry that captures the properties of flat space–time in 4 + 2 dimensions, including two times. Then by analyzing the hidden SO(4, 2) symmetry of the system *the observer can indirectly experience the various effects of the extra space and time.*

Do we know such SO(4, 2) symmetric 1T-physics systems in 3 + 1 dimensions? Yes, we do!

An SO(4, 2) symmetry known as "conformal symmetry" is widely recognized to be present in the equations that describe massless particles, and more generally for massless field theories in 3 + 1 dimensions. This corresponds to the shadow at the top in Fig. 7.7 indicated as "massless."

The famous physicist Paul Dirac proposed to explain SO(4, 2) conformal symmetry by associating it with 6 dimensions with constraints. His 1936 proposal was much later pursued by another famous physicist Abdus Salam in 1969, and by others. This higher dimensional approach to conformal symmetry was forgotten for nearly 30 years, until I rediscovered it independently in a different context in 1997 as a particular corner of 2T-physics. Dirac did not have a hint of the underlying Sp(2, *R*), nor of the multiple shadows . Now we understand that Dirac's approach to conformal symmetry, treated in the language of 2T-physics, corresponds to one specialized solution of the Sp(2, *R*) constraints in Fig. 7.6 that corresponds to the shadow marked "massless" in Fig. 7.7. This result applies in particle theories as well as in corresponding field theories.

Another lesser known case of hidden SO(4, 2) symmetry is the so-called dynamical symmetry of the hydrogen atom, which is shown at the bottom of Fig. 7.7. The SO(4) part of this symmetry is more familiar and has been known for a long time under the name of "Runge-Lenz vector," as being a "hidden" symmetry of the

---

[9] SO(4, 2) is the Lorentz symmetry in 4-space and 2-time dimensions. In the Cartan classification of symmetry groups outlined in footnote 1 of Chapter 5, it corresponds to an analytic continuation of D_3.

Hamiltonian of the H-atom. The rest of SO(4, 2) was noted occasionally as an algebraic property of the H-atom, but not recognized as a symmetry. However, through the formalism of 2T-physics, it was finally understood that the SO(4, 2) is actually a hidden *symmetry of the action* of the H-atom, just like conformal symmetry is a symmetry of the action, but not of the Hamiltonian. By being a hidden symmetry of the action, SO(4, 2) governs all properties of the H-atom as explained in detail in [16]. Some of those hidden properties provide a very interesting "window" to the 4 + 2 dimensions as discussed in the next section.

After the advent of 2T-physics the other surprising cases of hidden SO(4, 2) symmetries in every shadow in Fig. 7.7 became evident. No one suspected that massive particles in 3-space dimensions or the massive harmonic oscillator in 2-space dimensions have a hidden SO(4, 2) symmetry. The curved space marked as AdS_4 was known to have SO(3, 2) symmetry, but the larger hidden symmetry SO(4, 2) was not noted. The same applies to the other curved spaces marked as $AdS_3 \times S^1$, $AdS_2 \times S^2$, Robertson–Walker expanding universe, and all conformally flat space–times in 3 + 1 dimensions including certain singular (black hole type) cases. All of these have hidden SO(4, 2) symmetry, that is a larger symmetry than previously known for each of these examples [16, 18, 41].

Similarly, twistors were known to capture SO(4, 2) properties for massless systems in flat 3 + 1 space–time, but 2T-physics showed that the same twistors describe all the other cases in Fig. 7.7, thus discovering new systems where the twistor formalism can be usefully applied to describe the dynamics.

All of these SO(4, 2) hidden symmetries in 1T-physics systems are now recognized to be nothing but different 3 + 1 perspectives of the symmetries of flat space–time in 4 + 2 dimensions. These new perspectives provide various theoretical and experimental tools for 3 + 1 observers to experience indirectly the extra space and time.

The above properties can be re-stated for 2T-physics in d + 2 with an arbitrary number d of space dimensions plus 2 time dimensions. Then the emergent 1T-physics is in 1 less time and 1 less space dimensions. Any extra space dimensions and their possible relevance in nature can be accommodated by extending the system in the central and other blobs in Fig. 7.7 by adding more space dimensions as in string theory. The additional space dimensions would then be interpreted like those of Kaluza–Klein or string theory scenarios of tiny curled-up geometries as in Chapter 6, or like those discussed in the second part of this book of large extra dimensions in the scenario of Randall and Sundrum. Either case is accommodated in 2T-physics in $d + 2$ dimensions.

It is worth repeating here that all of the formal insights described above are conveyed by the informal shadows allegory described in Section 5.2, namely observers stuck on 2-dimensional walls see many shadows of the same object moving in a 3-dimensional room and think of the shadows as different "beasts" performing apparently unrelated motions. But, with enough research, they can in principle discover that the shadows are related, and in that way they can reconstruct the 3-dimensional object and its motion in the room.

To apply this allegory we replace the 3-dimensional room with the 2T-physics formulation of a system in a space–time with $d + 2$-dimensions. In the example of this section the *flat* d + 2-dimensional equations are just $X \cdot X = 0$, $P{\cdot}P = 0$, and $X{\cdot} P = 0$ as shown in Fig. 7.6. Similarly, the shadow "beasts" as well as the observers on the 2-dimensional walls are replaced by the shadow dynamical systems and the observers in $(d - 1) + 1$ dimensions as in Fig. 7.7. Then 2T-physics makes predictions that these 1T-physics systems should be related to one another by dualities and that specific hidden information about $d + 2$ dimensions should be found in each one of these 1T systems in $(d - 1) + 1$ dimensions.

Chapter 8
Evidence of 4 + 2 as Subtle Effects in 3 + 1 Dimensions

The observable effects of the *extra 1-space and 1-time* dimensions of 2T-physics can be found at all scales of physics, large and small, including quantum mechanical domains, in 3+1 dimensions. Often this comes methodically in the form of testable predictions of subtle hidden symmetries and hidden relations among dynamical systems that 1T-physics on its own misses to explain or predict, except for a few cases that were stumbled upon in the past. A couple of surprising examples in non-relativistic physics are discussed in this section, while the important case of conformal symmetry in relativistic field theory, along with some others related by dualities, will be discussed in the following section.

8.1 H-Atom

A very useful example of a subtle symmetry that has been known for a long time, but whose origin remained unexplained, is exhibited by some remarkable properties of the hydrogen atom.

As indicated in Fig. 7.7, the H-atom is one of the shadows that captures the properties of flat 4+2-dimensional phase space subject to the Sp(2, R) constraints. 2T-physics predicts that the H-atom must exhibit a remarkable hidden SO(4, 2) symmetry which is none other than the SO(4, 2) Lorentz symmetry of flat 4-space and 2-time dimensions, as seen from a special perspective.

Indeed, the H-atom has been known for a long time to have a hidden SO(4) symmetry of the Hamiltonian, as well as a "dynamical symmetry" SO(4, 2). A "dynamical symmetry" is not a symmetry, but rather it is an algebraic construct that cleverly facilitates computations of some properties of a physical system by using the mathematics of group theory. Until recently, it was not recognized that this dynamical symmetry for the H-atom exists precisely because the *action* of the H-atom (not its Hamiltonian) has an SO(4, 2) hidden symmetry. This was first understood through 2T-physics in 1998 [16]. Furthermore, the fact that this symmetry has been related to 4+2 dimensions provides a unique window for comprehending the effects produced in 3+1 dimensions by moving around in the extra dimensions as viewed from a special perspective in 3+1 dimensions, as explained below.

I. Bars, J. Terning, Founding Editor: F. Nekoogar, *Extra Dimensions in Space and Time*, Multiversal Journeys, DOI 10.1007/978-0-387-77638-5_8,

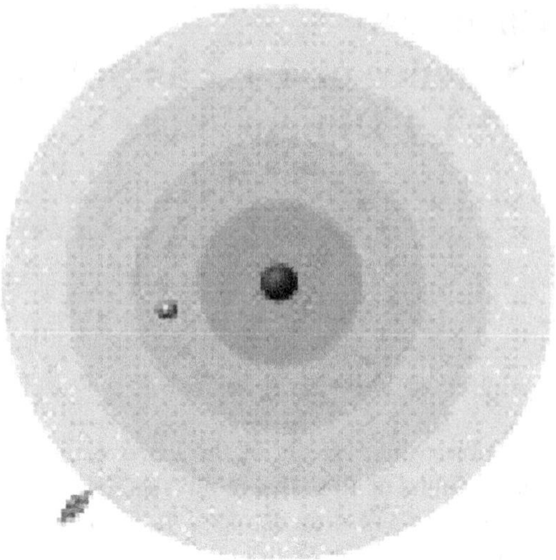

Fig. 8.1 Electron orbiting the nucleus. If it is in an excited energy state, it jumps to a lower energy state by emitting radiation. See footnote 1 (Credits, Physics-2000 website.)

The H-atom is the simplest atom, consisting of a single electron bound by the electromagnetic force to a nucleus made up of a single proton. The motion of the electron is somewhat analogous to the motion of the Earth orbiting the Sun as shown in Fig. 8.1, except for the additional rules of quantum mechanics that blurs the trajectory into a probability cloud where the electron can be found. Quantum mechanics explains that the electron has a unique lowest energy state in which it is stable and does not radiate energy.[1] In this lowest state the electron's average motion within a spherical probability cloud is along the radius, performing a "breathing" kind of in and out motion at any fixed angle, *without any rotation*.

This ground state is represented by the horizontal line segment at the bottom of the energy level plot in Fig. 8.2. The plot shows that the ground state has quantum numbers $n = 1$ and $l = 0$. The $n = 1$ refers to the lowest energy state, while $l = 0$ is the information that the angular momentum in this state is zero. There is no angular momentum because there is no rotational motion around the nucleus – the motion is purely radial. There is a single ground state, so on top of the horizontal segment the number 1 is written to indicate that there is only one such state with the quantum numbers $n = 1, l = 0$.

[1]Classical mechanics fails miserably. According to the rules of classical mechanics and classical electricity and magnetism, an accelerating charge must radiate electromagnetic energy, thus losing some of its own momentum and energy. An orbiting electron, as in Fig. 8.1, is accelerating since its velocity changes direction at every instant. Hence it must radiate and lose energy and momentum. Then like an orbiting satellite that is slowing down it should fall to the center of attraction. Estimates, according to classical mechanics, show that the electron would fall into the nucleus, so that the atom would be destroyed in 10^{-10} s. Evidently this is not true since atoms keep on surviving. This puzzle is resolved by quantum mechanics, which explains that the electron can lose energy by radiating photons if it is in an excited energy state. However, it cannot fall below the ground energy state where it is stable against radiation.

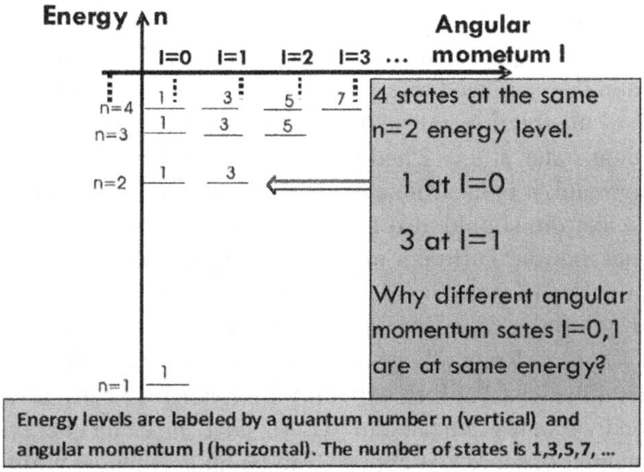

Fig. 8.2 "Seeing" 4+2 dimensions through the H-atom. Why motions with different amounts of angular momentum have the same energy (horizontal patterns at each n)? This is explained by rotation symmetry SO(4) between the 3 dimensions and the fourth space dimension. Also, the vertical ladder-like patterns of energy levels at each value of l are precise properties of the boost symmetry SO(1, 2) between the foruth space and the two time dimensions. The SO(4) and the SO(1, 2) are parts of the overall symmetry SO(4, 2).

At the next energy level labeled with $n = 2$ in Fig. 8.2 there are four quantum states. One of them has $l = 0$, while three of them have $l = 1$ as marked in the figure. When the atom is excited to a higher energy level, the electron orbits the nucleus in a probability cloud at a larger average radius, and after a while it makes a quantum jump to the lower energy level by emitting a photon as in Fig. 8.1. In the excited state, the $l = 0$ electron moves only radially as in the ground state described above. However, when it is in the $l = 1$ state, it rotates around the nucleus with a quantized amount of angular momentum denoted by $l = 1$.

There are three independent states at $l = 1$ because in 3-space dimensions the axis of rotation can be oriented along three independent orthogonal directions.

The relevant issue we wish to discuss is why do the 4 states at the $n = 2$ excited level have the same energy?[2] This will provide a subtle but powerful signal of the presence of a fourth space dimension.

---

[2]There is actually a very small energy difference between the $l = 0$ and $l = 1$ states. But it is effectively zero as compared to the other energy differences shown in the figure because it is smaller by a factor of 10^8. The actual energy difference is known as the Lamb shift, whose experimental measurement was worth a Nobel prize to Willis Lamb. The source of the interaction that splits these energy levels is found in higher order quantum corrections that can be understood and computed correctly in quantum electrodynamics (QED). The accuracy of this, and similar QED computations that are in agreement with experiment to 12 decimal places, is one of the most impressive triumphs of relativistic quantum field theory, showing the degree to which we have managed to understand nature. The fact that the Lamb shift is not zero does not change the conclusions of the hidden symmetry discussed in the text. It only means that in the presence of the quantum corrections the symmetry properties are slightly altered in 1T as well as in 2T-physics.

It is easy to understand why the three states at $l = 1$ have the same energy. Intuitively one might expect that all orientations of the axis of rotation produce the same energy, and this is indeed true according to the mathematical equations. Since the orientation of the rotation axis in three orthogonal dimensions cannot matter, the three independent states at $n = 2$ and $l = 1$ must have the same energy.

On the other hand, it is far from evident why the $l = 0$ state, with a purely radial motion of the electron, should also have the same energy as the $l = 1$ state with a purely rotational motion. Different motions are expected to result in different energies, why then in the H-atom the electron has the same energy for these states? In 1T-physics the reason is found in a hidden symmetry of the H-atom Hamiltonian known as the Laplace–Runge–Lenz–Fock–Pauli symmetry. This was explored by the Nobel prize physicist Pauli to explain the properties of the H-atom, including the fact that the four states indicated in Fig. 8.2 have the same energy. However, this symmetry remained obscure – it appeared to be an accidental symmetry without a deeper meaning or deeper origin.

2T-physics explains beautifully the origin and the meaning of the hidden symmetry. The allegory in the next paragraphs together with Fig. 8.3 is helpful to visualize this. The symmetry becomes evident as being simply the symmetry of rotations in 4-space dimensions. The three states with angular momentum $l = 1$ are associated with the three orthogonal directions, $\hat{x}$ or $\hat{y}$ or $\hat{z}$, in 3-space, while the fourth state is associated with the fourth space direction $\hat{w}$ that is orthogonal to the other three. The $\hat{x}$, $\hat{y}$, $\hat{z}$ states carry information about orientation in 3-space and therefore have angular momentum $l = 1$, but the $\hat{w}$ state has no information about orientation in 3-space and therefore has zero angular momentum in 3-space. However, since all orientations have the same energy in 4-dimensional flat space, and since the H-atom is a 3-dimensional shadow of a symmetric motion in four flat space dimensions (as in Fig. 7.7), all four states in Fig. 8.2 are predicted to have the same energy, and indeed they do!

It is illuminating to explain the role of the fourth dimension for the H-atom by using once more the shadows allegory. Consider motion in the room versus motion of the shadows on the wall as in Fig. 8.3. In this allegory the direction perpendicular to the wall labeled as w represents the "fourth dimension." Observers that are stuck on the wall do not have the privilege of being in the room, so they can see and measure only shadows on the wall.

The upper part of the figure illustrates a circular motion in the room of "electrons" rotating in an x–y plane parallel to the wall. The shadow of this motion on the wall clearly shows "electrons" rotating around a center, so observers on the wall interpret this shadow as a state of the system that has angular momentum. In the allegory, consider this shadow as the analog of the $l = 1$ state at $n = 2$ in Fig. 8.2.

Next consider the same system of rotating "electrons" in the room, but reorientated such that the motion takes place in the x–w plane as shown in the lower part of Fig. 8.3. Then the shadow on the wall performs only radial (breathing) motion, moving toward or away the center in the $\hat{x}$- direction as shown in the figure. The motion in the room could also be reoriented to the y–w plane or *any fixed plane* perpendicular to the wall. In all such cases the shadow is still a radial breathing motion

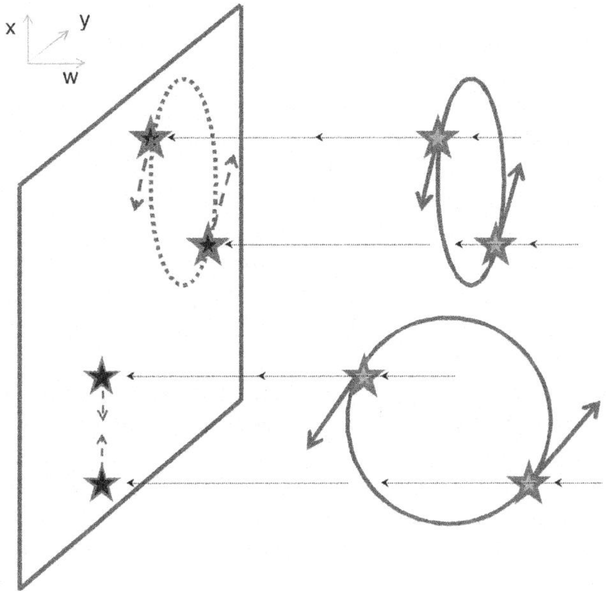

Fig. 8.3 Each electron in the room that orbits parallel to the wall (*upper part* of figure) produces a shadow that represents a rotating motion with angular momentum on the wall. But each electron in the room that orbits perpendicular to the wall (*lower part* of figure) produces radial motion, without any angular momentum, on the wall. However the energy of the motion in the room, and hence of the shadow, cannot depend on the orientation in the room

at *any fixed angle,* without any rotation. So observers on the wall interpret this as a state with zero angular momentum, which can occur with equal probability at any angle. To observers on the wall, this shadow is the analog of the $l = 0$ state at $n = 2$ in Fig. 8.2.

From the point of view of an observer in the room, it is evident that both types of states must have the same energy on the basis of rotation symmetry in the room, so the shadows must have the same energy. This symmetry is distorted and obscured for the observers on the wall, but they get to notice it indirectly by the subtle fact that the energies of states with different angular momenta are the same as in Fig. 8.2.

This phenomenon happens not only at the $n = 2$ level but also systematically at every energy level from $n = 1$ all the way to $n = \infty$, as implied by the vertical dots in Fig. 8.2. The states at a fixed energy level n are united together by being in the same angular momentum state in 4 dimensions, so they differ from each other only by their orientations in 4 dimensions.

All of this is precisely understood as a consequence of the evident SO(4) rotation symmetry in the 4-dimensional flat space. This evident symmetry gets distorted in the shadow into the hidden SO(4) symmetry of the H-atom as observed in 3-space dimensions.

There is more to be learned from the H-atom about the 4+2-dimensional space–time, including the two times, but this is less intuitive and requires some mathematics of group theory for SO(1, 2) [16]. In Fig. 8.2 one should notice the energy ladders at each value of angular momentum l. The energy differences from one rung of the ladder to the next, at a fixed value of l, corresponds to a pattern associated with the mathematics of SO(1, 2) symmetry. The pattern is called the "positive discrete series representation" of SO(1, 2). The SO(1, 2) symmetry unites the fourth dimension w with the 2 time dimensions t, t'; these coordinates (w, t, t') are connected to each other by "Lorentz boosts" in the (wt), (wt') planes and rotations between the two times in the (tt') plane. The (wt), (wt') boost transformations act vertically on each ladder in Fig. 8.2 and unite the states within the ladder at each value of angular momentum l.

The presence of these energy level patterns corresponds to not only SO(1, 2) symmetry but also to SO(3) $\times$ SO(1, 2) symmetry. This SO(3) is the rotation symmetry in the usual 3 dimensions (x, y, z) and is part of the SO(4) rotation symmetry including the fourth dimension w. So the fourth dimension is united with the 3-space dimensions by the SO(4) symmetry acting on (x, y, z, w), and it is also united with the 2-time dimensions by the SO(1, 2) symmetry acting on (w, t, t') as follows:

$$\overbrace{\underbrace{x,y,z}_{SO(3)}\ \underbrace{w,t,t'}_{SO(1,2)}}^{SO(4,2)} \quad \text{or} \quad \overbrace{\underbrace{x,y,z,w}_{SO(4)}\ \underbrace{t,t'}_{SO(2)}}^{SO(4,2)} .$$

In fact, there is a full SO(4, 2) hidden symmetry of the *action* of the H-atom in the 1T-physics formulation, which was not noticed before the advent of 2T-physics [16]. This is just the shadow of the SO(4, 2) symmetry of flat space–time as fully explained in the 2T-physics formulation. The SO(4, 2) which acts on the 6 dimensions (x, y, z, w, t, t') can be analyzed in terms of its sub-symmetries, such as SO(3) $\times$ SO(1, 2) or SO(4) $\times$ SO(2), which correspond to different ways of probing the H-atom as discussed above.

SO(4, 2) explains many properties of the H-atom as well as other shadows in Fig. 7.7. These follow from the fact that the H-atom corresponds mathematically to a unitary irreducible representation, called the "singleton," for the symmetry group SO(4, 2). All other shadows included in Fig. 7.7 correspond to the same singleton representation of the symmetry group SO(4, 2), but analyzed in different mathematical bases, corresponding to the perspectives that create the shadows. This simple mathematical and physical property underlies all of the duality relations[3] that exist between the 1T shadows in Fig. 7.7.

---

[3]The singleton representation is distinguished by the eigenvalues of the Casimir operators of SO(d, 2). As indicated in Fig. 7.7, the quadratic Casimir eigenvalue is $C_2 = 1 - \frac{d^2}{4}$, which for SO(4, 2) reduces to $C_2 = -3$. All other Casimir eigenvalues are also fixed in the singleton representation. It should be noted that the conformal shadow for massless relativistic particles, that has the conformal symmetry SO(4, 2), also has the same $C_2 = -3$ as can be verified for the case of the

8.2 The Fourth Space Dimension in Celestial Mechanics

Kepler was the first to realize that planets moving around the Sun trace trajectories in the form of ellipses. Newton explained convincingly that the trajectory could only be in the form of an ellipse by deriving this fact mathematically from his equations of motion for mechanics and universal gravity.

Thus, a bound point particle moving under an inverse-square central force must have a trajectory in the form of an ellipse. The center of attraction is the Sun shown as a small black circle in Fig. 8.4; this is the focus of the ellipse from which the position vectors $\vec{r}$ emanate. The points 1, 2, 3, 4 in the figure correspond to the positions of the Earth relative to the Sun at those times during the year when it becomes the middle of summer (1), fall (2), winter (3), and late spring (4). The changing distance to the sun at points 1, 2, 3, 4, combined with the slant of the axis of self-rotation of the Earth causes the changing seasonal temperatures we feel on the Earth.

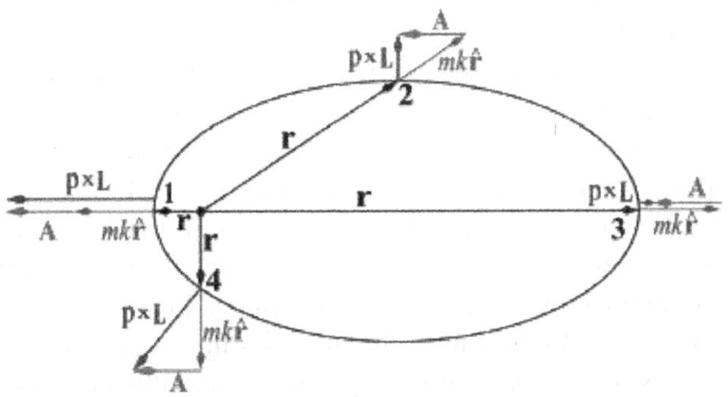

Fig. 8.4 The Laplace–Runge–Lenz vector $\vec{A}$, which represents the amount of rotation from usual space into the fourth dimension, is a constant at every instant during the motion of a planet. The direction and magnitude of $\vec{A}$ determines the orientation and the eccentricity of the ellipsoidal trajectory of the planet. This is information about the fourth dimension (Credits, Wikimedia Commons.)

The ellipse for a typical planet, such as the Earth, is non-changing over time: the ellipse sits on a plane that is stuck in the sky; furthermore, the orientation, or the axis, of the ellipse does not shift within the plane, so that the tip of the ellipse (called the perihelion) remains at the same point in the sky year after year.[4] What

---

massless Klein–Gordon field theory. This is not a coincidence, but is one small manisfetation of the expected duality between the massless particle and the H-atom, predicted by 2T-physics [16].

[4]It should be emphasized that unchanging ellipsoidal motion of planets is an approximation for an idealized model of non-relativistic motion in flat space–time. In real life, the perihelion of a planet does change slightly every year relative to other "fixed stars." There are some very small corrections in the equations that explain this fact, but these can be neglected for the sake of a very simple explanation of the main motion for an isolated Sun–planet system. The ellipse of a planet changes for several reasons. First, the Sun is actually moving within the Milky Way galaxy,

explains such a bizarre motion? Isn't it more intuitive to expect that the trajectory could change year after year?

The explanation of this phenomenon in 1T-physics involves two different symmetries of Newton's non-relativistic equations as applied to this special problem involving an inverse-square central force. The first is rotation symmetry in 3-dimensional space. Its consequence is that the angular momentum of the Sun–planet system cannot change over time – whatever its value is this year, it will remain the same every year at points 1, 2, 3, 4 in the figure, as well as at all instances of the planetary motion, past, present, or future. Angular momentum $\vec{L}$ (not shown in the figure) is a vector that does not change in magnitude or direction – its direction in Fig. 8.4 is into the plane at each point on the orbit.[5] Since the angular momentum vector must remain fixed in direction, the plane of the ellipse must also remain fixed in the sky. Hence, the underlying simple reason for the unchanging plane of motion is the *rotation symmetry* of the laws of motion.

But rotation symmetry in the usual 3-dimensional space is not enough to explain why the axis of the ellipse and hence the perihelion also remains fixed within the plane year after year (see footnote 4).

This is explained by another symmetry of the system governed by an inverse-square central force. There is a hidden symmetry of the equations of this special system that is harder to notice. The hidden symmetry results in a second vector $\vec{A}$, called the Laplace–Runge–Lenz vector, shown in Fig. 8.4. The vector $\vec{A}$ remains fixed in the sky in direction and magnitude, at each point of the trajectory. The axis of the ellipse is parallel to the constant vector $\vec{A}$, so this constant vector determines the constant orientation of the ellipse within the plane. Furthermore, the unchanging magnitude of the vector $\vec{A}$ determines the unchanging eccentricity of the ellipse. A large vector $\vec{A}$ leads to a large eccentricity, while a small vector $\vec{A}$ leads to a small eccentricity, with a zero vector $\vec{A}$ implying an exactly circular orbit.

---

so it drags the whole ellipse with it. Second there are other planets nearby whose effects can alter the trajectory of any given planet. Third, and more importantly, as discovered by Einstein, the perihelion of the ellipse for an *isolated* and idealized Sun–planet system does precess a tiny amount each year due to the modification of the gravitational force explained by general relativity. All of these very small corrections that are incorporated in 1T-physics are also present in 2T-physics. So the SO(4) symmetry discussed in the text, including the fourth dimension, is slightly altered, thus explaining correctly the exact observed motion. However, the conclusions on the fourth dimension arrived at through the simpler approximation remain just as valid after including all corrections.

[5] Angular momentum, which is given by the expression $\vec{L} = \vec{r}(t) \times \vec{p}(t)$, is a vector perpendicular to the plane formed by the two vectors $\vec{r}(t)$ and $\vec{p}(t)$. Here $\vec{r}(t)$ is the vector that represents the position of the planet as measured from the Sun. In Fig. 8.4 the vector $\vec{r}(t)$ is shown at four different locations, marked by 1, 2, 3, 4 during the course of the year. The vector $\vec{p}(t)$ (not shown in the figure) represents the momentum of the planet. At any instant $\vec{p}(t)$ is tangent to the ellipse and points in the direction of motion. Both $\vec{r}(t)$ and $\vec{p}(t)$ lie in the plane of motion. Both vectors $\vec{r}(t)$ and $\vec{p}(t)$ change as a function of time, as seen in particular for $\vec{r}(t)$ at the times when the planet is at locations 1, 2, 3, 4 in the figure. Even though both $\vec{r}(t)$ and $\vec{p}(t)$ change as a function of time, the angular momentum $\vec{L}$ computed as the cross product of these two vectors is always independent of time – it does not change in magnitude or direction – its direction in Fig. 8.4 is into the plane at each point on the orbit.

Basically, once the constant vectors $\vec{L}, \vec{A}$ are given at some instant, the rest of the motion, namely the changing position $\vec{r}(t)$ and momentum $\vec{p}(t)$ are tightly determined by the requirements that the two vectors $\vec{L}, \vec{A}$ cannot change as a function of time.[6] So, a key ingredient in planetary motion is two symmetries: rotation symmetry that leads to the constancy of $\vec{L}$ – which determines the plane, and the hidden symmetry that leads to the constancy of the Laplace–Runge–Lenz vector $\vec{A}$ – which determines the orientation and eccentricity of the ellipse within the plane.

Unlike rotation symmetry that has a clear geometrical meaning, the origin and significance of the hidden symmetry in this system remains obscure in 1T-physics. It happens to be an accidental symmetry which one can verify mathematically upon examination of the equations of motion for the inverse-square central force problem, but it has no clear physical meaning in 1T-physics.

Here is where 2T-physics brings new insight to this problem. The mathematical structure of the planet–Sun system is the same as the electron–nucleus system in the H-atom. Although the fundamental physics nature of the forces in these two cases is quite different, they lead to the same mathematics as the inverse-square central force problem in the non-relativistic Newtonian approximation. Therefore, just like the H-atom case, the celestial mechanics system can be regarded as a special shadow of the free motion in 4+2 dimensions subject to the Sp(2, R) constraints, as indicated in Fig. 7.7. Hence this system must have the hidden symmetry SO(4, 2), where the significance of SO(4, 2) is the Lorentz symmetry of 4+2 dimensions, which is distorted on the wall because of the perspective of the observer on the wall. The SO(4) part of this symmetry is rotation symmetry in 4 dimensions, and this explains directly the conservation of angular momentum $\vec{L}$ and of the Laplace–Runge–Lenz vector $\vec{A}$ in the 3-dimensional "wall."

In fact the x-, y-, z-components of the vector $\vec{L}$ can be described as angular momentum for rotations in the (xy), (yz), (zx) planes respectively. Similarly, the x-, y-, z-components of the vector $\vec{A}$ can be described as angular momentum for rotations in the (wx), (wy), (wz) planes, respectively. The constancy of both $\vec{L}, \vec{A}$ is just due to the rotation symmetry in 4 dimensions rather than only 3 dimensions.

To explain this more clearly we turn again to the shadows allegory as illustrated in Fig. 8.3, where the z-direction cannot be shown. So we consider the three directions x, y, w and the corresponding angular momentum for rotations in the (xy), (wx), (wy) planes. In this allegory ordinary space is the wall in the (xy) plane, so angular

[6]The vector $\vec{A}$ satisfies the equation $mk\hat{r}(t) + \vec{A} = \vec{p}(t) \times \vec{L}$ that is graphically represented in Fig. 8.4 at the four points 1, 2, 3, 4. As seen in the figure, adding the same constant vector $\vec{A}$ to the changing vector $mk\hat{r}(t)$ results precisely in the time dependent vector $\vec{p}(t) \times \vec{L}$. Recall that $\vec{L}$ is another constant vector perpendicular to the plane of the ellipse. Put another way, the vector $\vec{A}$, given by the expression $\vec{A} = \vec{p}(t) \times \vec{L} - mk\hat{r}(t)$, is constructed as the sum of two time-dependent vectors $\vec{p}(t) \times \vec{L}$ and $-mk\hat{r}(t)$. Here m is the mass of the planet and k is a constant that is related to Newton's gravitational constant, while $\hat{r}(t)$ is the direction of the vector $\vec{r}(t)$. The vector $\vec{p}(t) \times \vec{L}$ has a changing magnitude and direction as a function of time, while the vector $-mk\hat{r}(t)$ changes direction, but not magnitude as seen at the points 1, 2, 3, 4 in the figure. However, the magic of the hidden symmetry is that their sum remains a constant throughout the motion.

momentum $\vec{L}$ has only one component (the former z-component) corresponding to rotations in the (xy) plane, while the vector $\vec{A}$ has only the x-, y-components which correspond to angular momentum for rotations in the (wx), (wy) planes.

A subset of the motions in the "room" that is allowed by the constraints in the central blob of Fig. 7.7 corresponds to the shadow that behaves just like the planet–Sun system on the wall. This motion in the room is restricted to occur on any great circle of a perfectly round globe in the x, y, w space. Examples of such great circles are shown in Fig. 8.3. In these examples the circles in the room are oriented in two special directions. More generally they can be oriented in any direction. The diameter of the circle in the room (or the size of the globe) depends on the energy of the motion. The shadows of any such circles appear on the wall as ellipses of different sizes and different eccentricities. The eccentricity of the ellipse depends on the orientation of the circle in the room, and hence it is determined by the components of angular momentum for rotations in the (xy), (wx), (wy) planes, which means it is determined by the magnitudes of $\vec{L}$ and $\vec{A}$ and the orientation of $\vec{A}$.

For example the upper part in Fig. 8.3 shows a circle in the room in the (xy) plane and a corresponding shadow that is a perfect circle, so the "ellipse" on the wall degenerated to a circle. It has zero eccentricity since there is no rotation in the (wx), (wy) planes in the room. By contrast the lower part in Fig. 8.3 shows a perfect circle in the room in the (wx) plane and a corresponding "ellipse" that degenerated to a line segment on the wall, and hence it has extreme eccentricity.

This explanation clearly shows that the meaning of the Laplace–Runge–Lenz vector $\vec{A}$ is precisely angular momentum for rotations from ordinary space into the fourth dimension. The complicated mathematical expression for the vector $\vec{A} = \vec{p}(t) \times \vec{L} - mk\hat{r}(t)$ can actually be shown to be the shadow of a simple expression for angular momentum in the $(wx$, (wy), (wz) planes in 4-space dimensions [16]. The perspective of the shadow distorts and hides the underlying meaning that indicates the presence of a fourth space dimension.

In a similar way other shadows reveal other hidden relationships that correspond to other perspectives of motion in 4+2-dimensional space–time. So we do experience the effects of the second time dimension through the subtle behaviors of dynamical systems. We only need to be aware of how we interpret it. Once this is understood we learn that systems that appear to be different in ordinary space–time (as presented by 1T-physics) are actually related to each other, like shadows of the same object are related to each other. There are hidden symmetries in many systems that reflect the symmetries of the higher space–time. For example, the conformal symmetry that Dirac discussed is the hidden symmetry of one of the shadows – so Dirac's explanation of conformal symmetry is just one of many consequences of 2T-physics.

With the guidance provided by 2T-physics, the 1T observers, that are stuck on the "wall" in 3-space dimensions, can decipher the underlying subtle physics that has a simple significance in the larger space–time, and in this way "see" *indirectly* the many effects of the fourth space dimension and the second time dimension.

Chapter 9
Fundamental Universe as a Shadow from 4 + 2 Dimensions

Our current understanding of the deeper aspects of our universe in terms of the fundamental forces and fundamental matter was described in Chapter 2. As illustrated in Figs. 2.3 and 2.6, the smallest bits of matter that have been established experimentally are the quarks, leptons, and force particles, while their interactions are governed by the strong, weak, electromagnetic, and gravitational forces.

The gravitational force and its effect on all matter is formulated in detail as general relativity. The strong, weak, and electromagnetic forces and their effect on fundamental matter are formulated as a Yang–Mills gauge theory based on the symmetry group SU(3) × SU(2) × U(1) and is called the standard model of particles and forces. The mathematical structure of both of these theories is *local relativistic* field theory in 3-space and 1-time dimensions.

The standard model, *after taking into account quantum theory*, agrees with all experiments conducted so far all the way down to 10^{-18} m deep into the structure of matter. This is a distance scale of about 1000 times smaller than the proton. Up to now there is not a single experiment that contradicts the detailed predictions of the standard model, while in some special cases of very detailed measurements the accuracy of the agreement with theory has been proven up to 12 decimal places. This very impressive success shows that our theoretical understanding of physical phenomena involving the strong and electroweak interactions, at both conceptual and mathematical levels, is very deep indeed.

General relativity has also made numerous impressive predictions and passed all experimental tests. These successful tests are at macroscopic distances ranging from a few millimeters all the way to cosmological scales. But these tests involve weak to moderately strong gravitational fields in relatively large regions of space (outside of the atom) for which quantum effects are negligible. Therefore general relativity has so far been tested only at the classical level.

Quantum effects become significant at subatomic distances. At such small scales, the gravitational force, acting on elementary particles of very small masses, is negligible because it is extremely weak as compared to the strong and electroweak interactions acting on the same particles (see Fig. 9.1). However, gravity gets stronger and stronger at much smaller scales. It can be estimated that it does compete with the other forces acting on small particles in very special circumstances,

I. Bars, J. Terning, Founding Editor: F. Nekoogar, *Extra Dimensions in Space and Time*, Multiversal Journeys, DOI 10.1007/978-0-387-77638-5_9,

Property	Gravitational Interaction	Weak Interaction (Electroweak)	Electromagnetic Interaction	Strong Interaction
Acts on:	Mass – Energy	Flavor	Electric Charge	Color Charge
Particles experiencing:	All	Quarks, Leptons	Electrically Charged	Quarks, Gluons
Particles mediating:	Graviton (not yet observed)	W^+ W^- Z^0	γ	Gluons
Strength at 10^{-18} m	10^{-41}	0.8	1	25
Strength at 3×10^{-17} m	10^{-41}	10^{-4}	1	60

Fig. 9.1 Relative strengths of the strong and electroweak interactions versus gravity at about 1000th the size of the proton. Electromagnetic and weak interactions become comparable while the strong interaction is 25 times stronger, but gravity is weaker by a factor of 10^{-41} (Credits, Particle Data Group.)

such as at the Big Bang which involves interactions at the Planck scale.[1] This large gravitational force competes with the unified strong and electroweak force at the Planck scale as shown in Fig. 9.2, so they all contribute in shaping the universe at the time of the Big Bang. In this setting gravity acts strongly at very short distances where quantum effects are very significant. Therefore, formulating and understanding quantum gravity at the quantum level, like the other forces, is essential both as a matter of principle and for describing the physical phenomena correctly at the Big Bang. Quantizing gravity is a continuing effort whose most successful form is string theory.

The standard model and general relativity capture undoubtedly part of the truth about our universe. However there are many concrete puzzles, some based on experimental observations and some based on theoretical reasoning, which require explanation, but the answers cannot be found within the theoretical constructs of the standard model and general relativity. We know we must go beyond these theories but, despite brilliant ideas that involve string theory,

[1] The reason for the weakness of gravity inside the atom can be understood by estimating the gravitational force between two massive bodies, by using Newton's formula $F = -G\frac{m_1 m_2}{r^2}$. Here $G = 6.67259 \times 10^{-11} \mathrm{m^3/kg^1/s^2}$ is Newton's gravitational constant, m_1, m_2 are the masses of the two bodies, and r is the distance between them. For two protons of tiny masses $m_{\text{proton}} = 1.6726231 \times 10^{-27}$ kg within the nucleus, at a distance $r_{\text{nucleus}} = 10^{-15}$ m, the magnitude of the gravitational force is $|F_{\text{nucleus}}| \simeq 1.9 \times 10^{-34}$ N. For two up-quarks of mass $m_u \simeq 0.002$ GeV at 1000th the size of the proton, this becomes roughly $|F_{\text{inside proton}}| \simeq 8 \times 10^{-34}$ N. Compared to the electromagnetic force on the same quarks this is weaker by a factor of about 10^{-41} as indicated in Fig. 9.1, and therefore there is no chance of observing the effects of gravity within the atom or the nucleus at the current level of experimentation with small particles. However, at much smaller distances, such as at the Planck scale $r_{\text{Plank}} \simeq 1.61605 \times 10^{-35}$ m, the magnitude of the gravitational force between two quarks grows dramatically by a factor of 10^{40} to $|F_{\text{Planck}}| \simeq 10^{-5}$ N. At that point Newton's formula is no longer the correct description and should be replaced by general relativity with all of its consequences, so the naive estimate is not accurate. In any case, the main point is that at the Planck length, which is probed with energies of about 10^{19} GeV, the gravitational force becomes as important as the other forces as shown in Fig. 9.2.

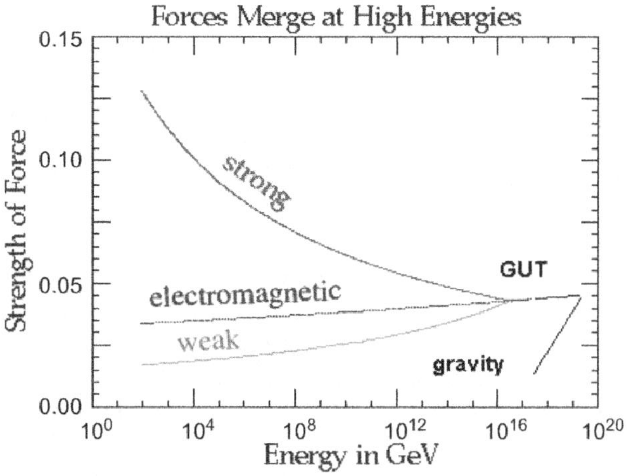

Fig. 9.2 At very short distances of the order of the Planck scale of 10^{-35} m, or at very high energies close to the Planck energy 10^{19} GeV, the force of gravity becomes as strong as the other forces (Credits, Particle Data Group.)

M-theory, and extra dimensions, the physically correct direction is yet to be established.

There are still missing conceptual ingredients. One of those missing ingredients within 1T-physics, which includes string theory and M-theory, is the fact that 1T-physics does not have the capacity to predict or explain the duality and hidden symmetry properties that were described in the previous chapters of this book. 2T-physics is the formalism that reveals those properties.

As will be described below, 2T-physics has managed to cast the standard model and general relativity in 3 + 1 dimensions as shadows of corresponding higher dimensional theories in 4 + 2 dimensions. So, 2T-physics describes our universe as we know it today as correctly as 1T-physics. This suggests that there exist deeper relationships, predicted by 2T-physics, that must be taken into account in the search of the deeper theory that explains the current puzzles. So 2T-physics should be included as one of the missing conceptual ingredients in the construction of a more fundamental theory.

The 2T-physics versions of the standard model, general relativity, and supersymmetric field theory in 4 + 2 dimensions are now described in the following sections.

9.1 The Standard Model as a 2T-Field Theory

One of the most commonly known symmetries in flat space–time in 3 + 1 dimensions is a "hidden" symmetry of massless particles called *conformal* SO(4, 2). Conformal symmetry governs some broad features of the dynamics of massless particles. This symmetry persists in many mathematical formalisms that describe massless particles, including particle dynamics on worldlines as well as classical field theory for massless fields including interactions. The classical conformal symmetry is known to persist at the quantum level in some special quantum field theory systems while it is broken by quantum effects in generic cases. Even if it is broken by quantum effects, conformal symmetry predicts definite observable consequences.

If we consider only the standard model in the absence of gravity[2] and in the limit of zero mass for the Higgs particle, then the standard model has *conformal symmetry* SO(4, 2) as a classical field theory. This symmetry, which can be verified within 1T-field theory, is a striking hint that the strong, weak, and electromagnetic interactions of all building blocks of matter – quarks, leptons and force particles – and their interactions could be described as shadows of their counterparts in 4 + 2 dimensions.

As indicated on the "massless" shadow in Fig. 7.7 conformal symmetry is a natural feature that emerges from 2T-physics in *flat space–time* in 4 + 2 dimensions. Like the hidden symmetries of the H-atom or celestial mechanics discussed in the previous section, the presence of conformal symmetry is a reflection of the presence of 1 extra space and 1 extra time dimensions in a flat space–time of 4 + 2 dimensions. So, from the point of view of 2T-physics, the effects of conformal symmetry in 1T-field theory must come from viewing a *2T-field theory* as a shadow from a special perspective.

That this is the case in field theory was established to some extent with the work of Dirac in 1936. This was independently rediscovered in 1998 in the context of 2T-physics as being the consequence of Sp(2, R) gauge symmetry in flat spacetime, and presented as being a property of the special *massless shadow*. In 2006 the field theoretic version of 2T-physics, that takes into account the effects of Sp(2, R) , was constructed [38], and in this field theory context, conformal symmetry has indeed been understood as a reflection of the underlying Lorentz-like symmetry SO(4, 2) in 4 + 2-dimensional flat space–time.

Therefore, we can interpret the well-known conformal symmetry as evidence for a higher space–time and use it as a probe for viewing and interpreting the higher dimensions from a special perspective. This is a complementary perspective as compared to the H-atom or celestial dynamic cases described in the previous section.

Following the construction of 2T-field theory, the formalism was immediately applied to building the standard model as a 2T-field theory. In this theory every degree of freedom that appears in the standard model – quarks, leptons, and force particles – is represented by a field in 4 + 2 dimensions. This means that the fields are taken as functions of space–time X^M in 4 + 2 dimensions and they carry vector

[2]When gravity is also included, the global SO(4, 2) symmetry is promoted to a local gauge symmetry in "tangent space". This is part of the 2T-physics approach to General Relativity. I should emphasize that 2T-gravity, or its 1T shadow, has no relation to what is known in the literature as the problematic "conformal gravity".

or spinor labels also in 4 + 2 dimensions to indicate their properties as spinning particles with various spins. For example the spin-1 photon field is represented by $A_M(X)$ where the vector spin index M on $A_M(X)$ and X^M runs over 4 + 2 dimensions. So the photon is described by the 6-component field $A_M(X)$ as compared to its shadow in 3 + 1 dimensions that has four vector components. Likewise, a neutrino in 4 + 2 dimensions is a chiral spinor of SO(4, 2) with four spinor components as compared to the shadow neutrino in 3 + 1 dimension which is a chiral spinor of SO(3, 1) with two spinor components. Similar comments apply to all gauge bosons (force particles) and fermions (matter particles) in 2T field theory.

At the naive level, compared to 3 + 1-dimensional fields the 2T-field theory appears to contain many more degrees of freedom. The extra degrees of freedom come from the extra fields needed to describe spin (6 instead of 4 for the photon, etc.) as well as from expanding the field in terms of an infinite tower of Kaluza–Klein-type modes. If this naive approach were the case, then the theory would inescapably contain ghosts due to the extra time, and it would have to be discarded on the basis of insurmountable mathematical and physical problems.

However 2T-field theory differs in its structure from a naive extension of 1T-field theory in ways not discussed before in field theory. The new structure follows from imposing the Sp(2, R) constraints of Fig. 7.6 which came from gauge invariance under the Sp(2, R) gauge symmetry in the worldline description of free particles. These constraints are incorporated into 2T-field theory by a procedure called "BRST field theory" and generalized in the presence of field interactions [37]. The field theory structure that emerges comes with new gauge symmetries not discussed before in previous naive approaches.

The role of the new gauge symmetry in 2T-field theory is twofold. First, as for other gauge symmetries, it imposes an overall unique structure that distinguishes it from generic field theories. Second, the gauge symmetry is just enough to remove the extra degrees of freedom and make the 4 + 2 field theory equivalent to a shadow field theory in 3 + 1 dimensions. In particular, the gauge symmetries remove all the ghost problems or causality problems that would have caused headaches in a generic 2T-field theory.

The shadows obtained from 2T-field theory are the same ones discovered for the particles on worldlines as displayed in Fig. 7.7, but now the language to describe them is 1T-field theory and it includes field interactions. The shadow 1T fields have the same number of degrees of freedom as one would expect in a 3 + 1 field theory namely there are no Kaluza–Klein degrees of freedom nor extra fields coming from the extended indices on fields.

Thus, 2T-field theory is a viable formulation to describe physical phenomena, either as a collection of gauge fixed shadows in 3 + 1 dimensions related to each other by dualities or as a unifying field theory directly in 4 + 2 dimensions that is the parent theory of all the 3 + 1-dimensional shadows.

When this formalism is applied to the 2T-field-theoretic formulation of the standard model in 4 + 2 dimensions, one finds that the *conformal shadow of the 2T standard model* produces the usual standard model as a 1T relativistic field theory in 3 + 1 dimensions. The conformal shadow is the shadow that corresponds to the "massless particle" in Fig. 7.7. This shadow automatically has the SO(4, 2) conformal symmetry expected for massless classical field theory in 3 + 1 dimensions. Evidently, in this formulation the meaning of this hidden conformal symmetry is the Lorentz symmetry SO(4, 2) of the larger space–time and is a probe for the extra dimensions as described above.

There are some new predictions about the standard model coming from this 2T formulation. The shadow standard model is basically the same as the familiar one in most respects, but it does come with some additional restrictions on the Higgs field sector that are not necessarily required in the usual 1T-field-theoretic formulation. The extra conditions are compatible with everything we know today experimentally about the standard model, but these conditions may become a way of testing 2T-physics in the future.

The modified structure of the shadow standard model leads to new twists on some little understood fundamental physical processes. In particular, in order to initiate the electroweak phase transition that explains the source of mass for all massive matter, the 4 + 2 standard model requires the presence of a *dilaton field* coupled to the Higgs field. The dilaton is required also by the 2T formulation of general relativity in 4 + 2 dimensions as discussed below. In the shadow standard model, the dilaton is responsible for inducing the Higgs to undergo the electroweak phase transition and in this way tie this phase transition to other cosmic phase transitions that occur during the evolution of the universe.

The extra conditions that lead to the Higgs-dilaton sector may also have implications on the resolution of some unsettled issues with the standard model, namely the "hierarchy problem" and the "strong CP violation problem." These aspects of the shadow standard model bring new conceptual guidance and new ways of thinking about unanswered questions in fundamental physics. When these issues are better clarified both experimentally and theoretically, such characteristic properties of the 2T standard model will help distinguish 2T-physics from other approaches.

9.2 2T-Gravity in 4 + 2 Dimensions

Having laid down the principles of 2T-field theory and the 2T standard model based on the Sp(2, R) constraints of flat space, $X^2 = P^2 = X \cdot P = 0$ in Fig. 7.6, the way was paved for extending 2T-field theory from d + 2-dimensional *flat* space–time to *curved* d + 2-dimensional space–time. The key for doing this was to first understand how to modify the Sp(2, R) constraints from flat to curved space. The appropriate modifications were already obtained in the years 2000–2001 as part of the 2T-physics worldline formalism by considering the phase space $X^M(\tau), P_M(\tau)$ of a 2T particle moving in the presence of gravitational, electromagnetic, or other more general backgrounds [23, 24]. In the case of pure gravity, the Sp(2, R) constraints featured explicitly the metric $G_{MN}(X)$ along with another field $W(X)$ that were required to satisfy certain equations, called homothety conditions, which signified that the curved space geometry in d + 2 dimensions has to have some special properties.[3]

The challenge was to find a field theory action in which, in addition to Einstein's field equations in d + 2 dimensions, these generalized homothety conditions on the metric $G_{\mathrm{MN}}(X)$ and $W(X)$ are derived from the same *action principle*. The desired action principle was finally achieved in 2008 [43]: for pure gravity the fundamental fields include not only $G_{\mathrm{MN}}(X)$ and $W(X)$ but also the *dilaton* field $\Omega(X)$. This triplet of fields is intertwined to each other in an action functional that features new gauge symmetries that extend the gauge symmetries of flat 2T-field theory in some remarkable new forms, consistently with Einstein's principle of symmetry under general coordinate transformations. These gauge symmetries permit only a unique form of the action, thus making 2T-gravity a theory completely determined purely by a special gauge symmetry.

Enlarging the theory, matter interacting with the gravitational field is included in the form of the 2T standard model coupled to gravity. There is a well-known geometrical approach for coupling the gravitational field $G_{\mathrm{MN}}(X)$ to matter fields to describe correctly their interactions. However, with the additional gauge symmetries of 2T-gravity, this coupling must obey certain additional rules that are more restrictive in comparison to 1T-field theory. The new features mainly involve some special couplings of matter fields to the dilaton $\Omega(X)$ and to $W(X)$.

After the dust settles, it turns out that the complete 2T-gravity action has no dimensional constants. At first sight this appears to deviate from standard approaches to gravity. So, by virtue of the stronger gauge symmetries, Newton's gravitational constant $G = 6.67259 \times 10^{-11}\ \mathrm{m^3/kg^1/s^2}$, which appears in Einstein's general relativity in the form $\frac{1}{16\pi G}$, is not permitted to appear in 2T-gravity because it is not a dimensionless pure constant. Instead of the constant $\frac{1}{16\pi G}$, a function of space–time, which may be called the "gravitational function," couples to the curvature. That function is $(\Omega(X))^2 - a\sum_i (S_i(X))^2$. It has a special quadratic form involving a sum over the dilaton $\Omega(X)$ and other scalar fields $S_i(X)$ analogous to the Higgs field, with $a = \frac{1}{12}$ for 4 + 2 dimension, or $a = \frac{d-2}{8(d-1)}$ for $d + 2$ dimensions, as enforced by the gauge symmetries.

The Higgs-type fields $S_i(X)$ drive cosmic phase transitions, such as inflation, grand unification, and electroweak phase transitions, that the universe as a whole undergoes as it cools down from the time of the Big Bang to the present.

[3] In flat space the Sp(2, R) constraints discussed in Fig. 7.6 were

$$X \cdot X = 0, X \cdot P = 0, P \cdot P = 0$$

In curved space, in the presence of gravity, these constraints get modified to

$$W(X) = 0,\ V^M(X)\,P_M = 0,\ G^{\mathrm{MN}}(X)\,P_M P_N = 0.$$

Here $V^M(X)$ is given by $V^M(X) = G^{\mathrm{MN}}(X)\frac{\partial W(X)}{\partial X^N}$, while the background fields $W(X)$ and the metric $G^{MN}(X)$ are required to satisfy certain geometric properties called "homothety conditions," so that the Lie algebra for the Sp(2, R) gauge symmetry is satisfied.

Having a function instead of the gravitational constant $\frac{1}{16\pi G}$ implies that, in different regions of space–time X^M, gravity can act with different strengths that depend not only on the curvature of space–time in that region but also on the value of the "gravitational function" $G(X)$ given by

$$\frac{1}{16\pi G(X)} = (\Omega(X))^2 - a\sum_i (S_i(X))^2 .$$

This appears to deviate from standard gravity, so can it be correct? No worries . . . we will discuss a beautiful effect of the gauge symmetries that shows how this theory agrees with the tested aspects of general relativity while also making some additional subtle predictions on the history of our universe and other physical processes in the present universe.

9.3 Dilaton-Driven Cosmic Phase Transitions

So how does 2T-gravity in 4 + 2 dimensions yield general relativity in 3 + 1 dimensions, with the usual gravitational constant $\frac{1}{16\pi G}$? The mechanism that explains this fact in 2T-physics is also the driving force for the emergence of other dimensionful constants in the historical evolution of the universe through cosmic phase transitions. The latter are Higgs -type mechanisms that generate masses of fundamental constituents, such as quarks, leptons, $W^{\pm}$, Z^0, Higgs particles, and others.

To understand the emergent fundamental physics for observers in 3 + 1 dimensions we must consider one of the shadows of 4 + 2 2T-gravity coupled to the standard model. As explained in Section 9.1, the appropriate shadow that resembles the usual relativistic field theory in 3 + 1 dimensions is the *conformal shadow* corresponding to the "massless particle" indicated in Fig. 7.7. In this setting the shadow of the dilaton and Higgs-type scalars $\Omega(X)$, $S_i(X)$ are fields $\phi(x)$,$s_i(x)$ in 3 + 1 dimensions x^{μ}, while the shadow of the gravitational function becomes $(\phi(x))^2 - a\sum_i (s_i(x))^2$.

In the presence of gravity the conformal shadow has another remarkable local symmetry which is called the Weyl gauge symmetry in 3 + 1 dimensions. The Weyl symmetry is actually the very first gauge symmetry proposed in the history of symmetries in physics. It is a symmetry that says the laws of physics remain the same under *local* dilatations or contractions of an observer's coordinate system. The original theory does not have a Weyl symmetry in 4 + 2 dimensions. The Weyl symmetry in 3 + 1 dimensions emerges from the general coordinate transformations in the extra 1-time and 1-space dimensions [45]. It is a property of the conformal shadow, but not necessarily of other shadows.

With such a Weyl symmetry the unit of length in space–time can be changed arbitrarily without affecting the laws of physics. It can be fixed to whatever units of length an observer wishes to choose. Once a unit of length is fixed, the Weyl symmetry appears to be broken for that particular observer's coordinate system. That observer measures everything in terms of the unit he defines, but the laws of physics are the same whether or not units are chosen for this or other observers that may choose a different set of units.

The Weyl gauge symmetry can be used to fix some definite units for us as 3 + 1-dimensional observers of the universe. One way of doing this is to fix the shadow of the dilaton $\phi(x)$ to be a constant everywhere in our universe, thus replacing the function $\phi(x)$ by a fixed constant ϕ_0. By not depending on space–time x^μ, the constant dilaton ϕ_0 has the same value for all observers everywhere in the universe for all times. This fixes the units of length and time for all observers in our universe.

With this choice of units let us examine the gravitational function where the constant ϕ_0 appears as $(\phi_0)^2 - a\sum_i (s_i(x))^2$. As explained below, one finds that this function changes *in a stepwise fashion* as the universe expands according to Fig. 9.3.

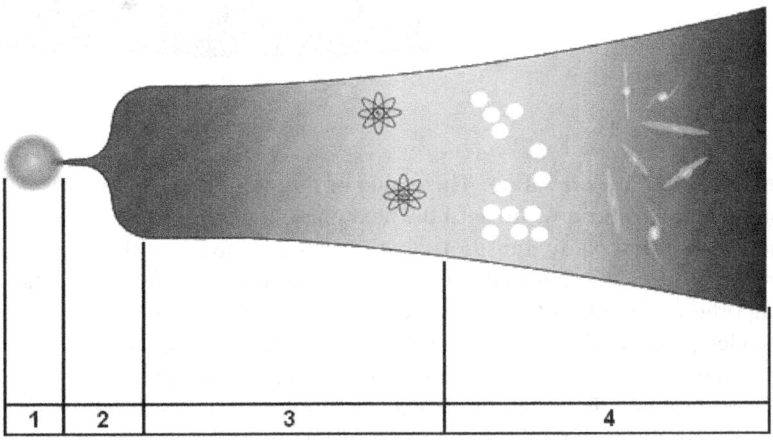

Fig. 9.3 Eras: (1) Big Bang, (2) inflation, (3) elementary matter, (4) galaxies (Credits, Particle Data Group.)

In the Big Bang phase (1) the value of the constant ϕ_0 dominates by far the value of the other fields $s_1(x), s_2(x), s_3(x), \ldots$, so that the gravitational function $(\phi_0)^2 - a\sum_i (s_i(x))^2$ is approximately the constant $(\phi_0)^2$. By writing this constant in the form $(\phi_0)^2 = \frac{1}{16\pi G_{BB}}$, we can identify G_{BB} as the gravitational constant during the Big Bang period (1) in Fig. 9.3. This G_{BB} is not Newton's constant $G = 6.67259 \times 10^{-11}\text{m}^3/\text{kg}^1/\text{s}^2$. We don't really know the history of the universe sufficiently well to determine the value of G_{BB} or of ϕ_0.

In the next phase (2), inflation of the universe is driven by the scalar $s_1(x)$, called the *inflaton*, which changes drastically its value by exponentially large amounts. At the end of inflation, $s_1(x)$ can be written in the form $s_1(x) = v_1 + \delta s_1(x)$, where v_1 is a large constant value while $\delta s_1(x)$ as well as $s_2(x), s_3(x), \ldots$ are all negligibly small. Therefore, at the end of inflation, the gravitational function $(\phi_0)^2 - a\sum_i (s_i(x))^2$ is approximated by the constant $(\phi_0)^2 - a(v_1)^2$. By writing

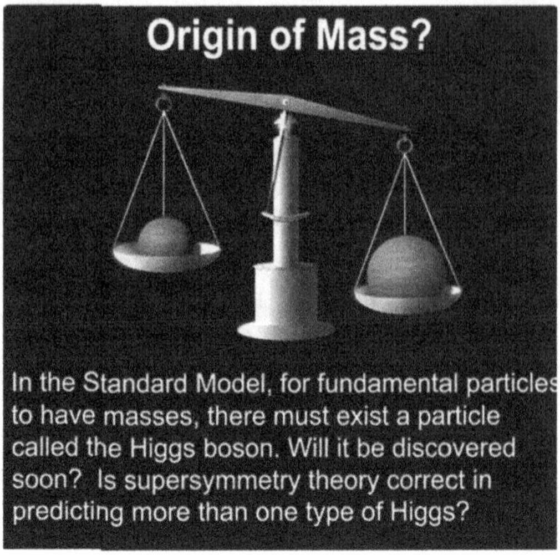

Fig. 9.4 The simplest Higgs scenario may be right. But several other possibilities are also permitted by our current understanding of the fundamental theory (Credits, Particle Data Group.)

this constant in the form $(\phi_0)^2 - a\,(v_1)^2 = \frac{1}{16\pi G_{\rm inf}}$ we can identify G_{inf} with the gravitational constant after inflation. This value of the gravitational constant describes gravitational forces for a while until the next phase transition.

There are several phase transitions that are believed to occur that are not explicitly shown in Fig. 9.3, but happen during the period (3) indicated in the figure. The important ones are the *grand* unification (GUT), the supersymmetry (SUSY), and the electroweak (EW) phase transitions. After the GUT occurs the gravitational function is dominated by the constant $(\phi_0)^2 - a\,(v_1)^2 - a\,(v_2)^2 = \frac{1}{16\pi G_{GUT}}$, where G_{GUT} is interpreted as the gravitational constant after the GUT phase transition. Similarly, after the SUSY or EW occurs the gravitational function is dominated by the constant $(\phi_0)^2 - a\,(v_1)^2 - a\,(v_2)^2 - a\,(v_3)^2 = \frac{1}{16\pi G_{EW}}$, where G_{EW} is interpreted as the gravitational constant after the SUSY or EW phase transition (the EW phase transitions is thought to be described by the Higgs particle, as in Fig. 9.4). There may be some other phase transitions between GUT and EW scales. Our awareness of the cosmic phase transitions could change as we understand better the physics beyond the standard model.

The important concept in the previous paragraph is that, according to the consequences of 2T-physics, *the gravitational constant may not be really a constant.* Rather, it could change in steps as a function of time during the cosmic evolution. In between cosmic phase transitions it is approximately a constant, but after each phase transition it changes to a new almost constant value. The current value of the gravitational constant G is approximately the value at the end of period 3 in Fig. 9.3. This is then the quantity $G_{EW} = G = 6.67259 \times 10^{-11}\text{m}^3/\text{kg}^1/\text{s}^2$ that was achieved when the universe was still very young, about 10^{-8} s old, when the electroweak phase transition is estimated to occur. Actually we know that the electroweak scale v_3 is smaller compared to the GUT scale v_2 by a factor of 10^{-14}, and similarly the GUT scale v_3 is smaller compared to the inflation scale v_1. In view of this, the EW transition can be neglected for the purpose of this discussion. So, one might estimate that the value $G = 6.67259 \times 10^{-11}\text{m}^3/\text{kg}^1/\text{s}^2$ is almost achieved already at the GUT transition, which is when the universe was roughly 10^{-21} s old, or even well before that, but certainly after inflation.

The non-constancy of the gravitational constant has implications for understanding the very early stages of the universe. Virtually all previous analyses are based on the 1T version of general relativity in which the gravitational constant is assumed to be a constant as an input in the theory. However, on the basis of quantum gravity we already have doubts that traditional general relativity may not be trusted at the time of the Big Bang or the time of inflation. So there is plenty of room for the version of gravity discussed above, which is a necessary consequence of 2T-physics. Better future understanding of the origins of the universe could provide a means of testing the scenario above predicted by 2T-physics.

The detailed consequences for cosmology of the above concept of a changing gravitational constant is under investigation at the time of this writing.

9.4 Electroweak Phase Transition

The electroweak phase transition (EW) is a crucial ingredient for understanding the origin of mass for all fundamental particles, including quarks, leptons, and force particles . The measured masses shown in Figs. 9.5 and 9.6 and an additional set of parameters called *mixing angles* span a bewildering pattern that remains unexplained within the standard model, but could possibly be determined by a unified

FERMIONS matter constituents spin = 1/2, 3/2, 5/2, ...

Leptons spin =1/2			**Quarks** spin =1/2		
Flavor	Mass GeV/c^2	Electric charge	Flavor	Approx. Mass GeV/c^2	Electric charge
ν_L lightest neutrino*	$(0\text{–}0.13)\times10^{-9}$	0	u up	0.002	2/3
e electron	0.000511	−1	d down	0.005	−1/3
ν_M middle neutrino*	$(0.009\text{–}0.13)\times10^{-9}$	0	c charm	1.3	2/3
μ muon	0.106	−1	s strange	0.1	−1/3
ν_H heaviest neutrino*	$(0.04\text{–}0.14)\times10^{-9}$	0	t top	173	2/3
τ tau	1.777	−1	b bottom	4.2	−1/3

Fig. 9.5 Masses and charges of quarks and leptons (Credits, Particle Data Group.)

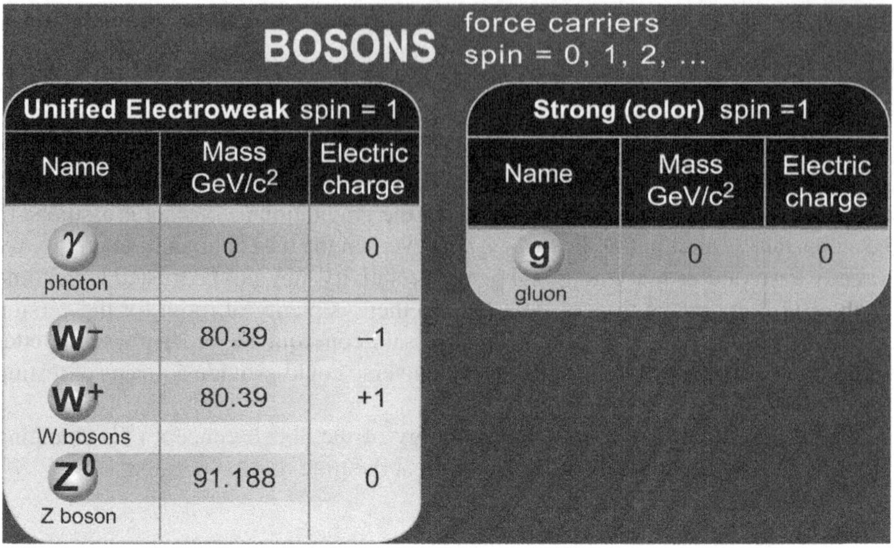

BOSONS force carriers spin = 0, 1, 2, ...

Unified Electroweak spin = 1		
Name	Mass GeV/c^2	Electric charge
γ photon	0	0
W$^-$	80.39	−1
W$^+$ W bosons	80.39	+1
Z^0 Z boson	91.188	0

Strong (color) spin =1		
Name	Mass GeV/c^2	Electric charge
g gluon	0	0

Fig. 9.6 Masses and charges of force particles (Credits, Particle Data Group.)

theory yet to be convincingly constructed. Understanding these masses and mixing angles is a major theoretical challenge for theories beyond the standard model. The resolution could open a floodgate to deciphering many other secrets of nature.

The precise details of the EW transition are still obscure, but it is strongly believed that a so far hypothetical particle, called the Higgs , is responsible for the EW phase transition. More complicated mechanisms that imitate the Higgs could also be at play, so that the Higgs particle may be only the tip of an iceberg as pointed out in Fig. 2.5.

Is the picture presented by the simplest version of the standard model correct or is there much more? It is now timely to discuss the various theoretical possibilities underlying the fundamental physics related to the EW transition because during 2009–2011 the details of the Higgs or of its underlying physics will be clarified through experiments to be conducted at the Large Hadron Collider (LHC) at the international laboratory CERN.

An intriguing conceptual issue that is not much discussed about the EW transition includes the following. In the usual 1T-physics approach to the standard model, the dilaton, which really is part of gravity, need not participate at all to explain the electroweak phase transition. This is because the usual standard model already contains the mass of the Higgs as a dimensionful parameter introduced "by hand." So the usual standard model is not scale invariant and does not need the dilaton to generate a scale. If one believes that the electroweak interactions are described in this way, without the participation of gravity or a dilaton, then a very strange situation arises: Without any reference to the history or the size of the universe, the electroweak phase transition occurs by suddenly filling up the entire universe by the constant value of the Higgs field. This is not believable. Phase transitions typically occur in very small regions of space–time, therefore it is very unlikely that an entire big universe can undergo a sudden and global phase transition. Hence this 1T-physics version of the electroweak phase transition is at best incomplete.

Let us contrast this with 2T-physics in which the dilaton must participate to drive the EW phase transition. Without going into details, it is important to emphasize that a very important consequence of the 2T version of gravity and standard model is that phase transitions cannot occur in the universe without the participation of the dilaton. This is because, in the necessarily Weyl scale-invariant shadow theory, scales are generated only by gauge fixing the dilaton as described in the previous section. So in the 2T-physics approach, the standard model necessarily includes gravity and the dilaton, in order to explain the cosmic phase transitions. This is quite welcome conceptually because then the standard model does not stand isolated from the evolution of the universe. The seeds for the EW phase transition were planted by the dilaton while the universe was still small. Then it is not unlikely for the entire, but small, universe to evolve to the state filled by the constant value of the Higgs field after the phase transition. Therefore the conceptual problem of the previous paragraph is naturally avoided in 2T-physics.

The dilaton-driven electroweak phase transition scenario may also resolve two other problems of the standard model. These are the "strong CP violation problem" and the "mass hierarchy problem." These are still under investigation in the 2T-physics framework, as a quantum field theory, and will not be elaborated here.

Chapter 10
Current Status of 2T-Physics and Future Directions

2T-physics has come a long way since its initial steps in 1995 [3, 4]. By now it has been established that nature obeys a fundamental general principle of broad significance. It can be stated as follows:

Sp(2, R) gauge symmetry of phase space is a fundamental property of nature.

All aspects of 2T-physics are outcomes of this principle which was formulated in 1998 [15]. The two-time feature is not an input but rather an output of the requirement of Sp(2, R) gauge symmetry. The success of 2T-physics in describing nature correctly from cosmological to subnuclear scales of physics is ample confirmation that this is indeed a correct principle.

It is remarkable that this principle has many testable consequences that *1T-physics on its own was unable to predict but was only able to confirm.* The Sp(2, R) principle enhances the traditional formulation of physics and should be included systematically in the more comprehensive formulation of all physical phenomena at all scales. 2T-physics is a formalism which is still evolving today and has not yet reached its final version.

The early development of 2T-physics was at the level of particle dynamics in a worldline formalism. In recent years the more comprehensive field theory formulation emerged:

2T-field theory in $d + 2$ dimensions compatible with the Sp(2, R) principle, and free of ghosts, has been constructed and successfully applied.

The 2T-field theory approach, developed since 2000 [25], evolved to its present form in 2006 after using a technique called BRST field theory [37]. It was then applied in $4+2$ dimensions to obtain the 2T standard model in 2006 [38] and the 2T formulation of gravity in 2008 [43, 45]. These have correctly reproduced the experimentally successful 1T standard model and 1T general relativity as their conformal shadows in $3+1$ dimensions. This leaves no doubt that 2T-physics is a correct and a complete description of nature in its most fundamental form as we know it today.

I. Bars, J. Terning, Founding Editor: F. Nekoogar, *Extra Dimensions in Space and Time*, Multiversal Journeys, DOI 10.1007/978-0-387-77638-5_10,

It is worth briefly summarizing again the relation between the $4+2$ and $3+1$ field theories. After a process called "gauge fixing" and solving some kinematic equations of motion, the $4+2$ field theory yields various "shadows" in $3+1$ dimensions. One of them is called the 1T *conformal shadow* of the 2T-field theory. It corresponds to the "massless particle" shadow in the simplest particle dynamics example shown in Fig. 7.7. Generally, the conformal shadow of the $4+2$ field theories coincides with the familiar 1T-field theories that agree with experiments, except for some additional new constraints. These new constraints on 1T-field theory – in particular on scalar fields and their interactions – are consistent with everything we managed to measure in experiments so far. Potentially there are measurable consequences of these new restrictions *within the conformal shadow* that could distinguish 2T-physics from other less restrictive 1T-approaches, as explained in the previous section.

Beyond the standard model and general relativity, the 2T-field theory approach is applied also to grand unified theories (GUTs) as well as to supersymmetric extensions of the standard model.

> Grand unified theories and supersymmetric 2T-field theory with $N = 1, 2, 4$ supersymmetries have been constructed as2T-field theories.

Here $N = 1, 2, 4$ refers to the number of supersymmetries, while $N = 0$ means no supersymmetry at all. For higher N the theory becomes more restrictive, and $N = 4$ is the largest amount of supersymmetry attainable in the absence of gravity.

Supersymmetry is a symmetry between bosons and fermions. It has some nice theoretical properties that make field theory more pleasing and more manageable mathematically. It has been postulated to be a property of nature to explain why it is plausible to have a broad range of masses in the same field theory without also having to include an even bigger mass range that extends all the way to the Planck mass. This is used in the supersymmetric version of the standard model to explain why the mass of the Higgs particle – expected to be several hundred giga electron Volts – is not as large as the Planck mass of 10^{19} GeV.

There may be other ways to circumvent this so-called mass hierarchy problem, so supersymmetry is not absolutely necessary. But due to its additional attractive mathematical features, and also because it is a necessary feature in string theory, a majority of particle physicists tend to believe that supersymmetry may be a property of nature at some energy level. Some physicists hope that the supersymmetry scale may be as low as the energies attainable by the Large Hadron Collider (LHC) at CERN. If this is the case, there is a chance of seeing the effects of supersymmetry during 2009–2011. Supersymmetry would manifest itself most clearly with the discovery of new particles that would be the supersymmetric partners of all the known quarks, leptons, and force particles, with definite predicted properties.

Supersymmetry in $4+2$ dimensions in the context of 2T-field theory has some parallels to supersymmetry in $3+1$ dimensions but it cannot be just a naive mathematical generalization because of the additional mathematical structures in 2T-field theory imposed by Sp(2, R). The technical aspects of this problem have been solved, and 2T-field theories with $N = 1,2,4$ supersymmetries have been constructed [39, 40, 44].

The supersymmetric 2T formulation of the standard model for $N = 1$ yields a supersymmetric conformal shadow in $3+1$ dimensions. The $N = 1$ case is the most likely form of the supersymmetric standard model that can fit the experiment if supersymmetry is discovered at the LHC. As in other cases, the conformal shadow comes with more restrictions as compared to the usual $N = 1$ supersymmetric standard model that obeys only the rules of 1T-field theory. Therefore, while most of the observable predictions are similar to the usual 1T supersymmetric model, there are a few differences in particular in the scalar particle sector. If supersymmetry is discovered at the LHC it would be of interest to explore if the differences can be tested.

The $N = 4$ supersymmetric field theory has some remarkable mathematical properties that make it more manageable than the less supersymmetric cases. So $N = 4$ is used as a powerful theoretical laboratory to compute predictions of field theory, and then extrapolate to the more realistic $N = 0$, 1 cases. For $N = 4$ 2T-field theory, the conformal shadow of the $4+2$ theory is exactly the $3+1$ theory without any deviations from its 1T counterpart. This connects directly with the fact that the $3+1$ theory is exactly SO(4, 2) conformally invariant even at the quantum level. This is one of the very important properties of the $N = 4$ field theory that has been used crucially in many of the ongoing theoretical investigations in $3+1$ dimensions that are still of intense interest today. Recall that SO(4, 2) is a direct consequence of $4+2$ dimensions as claimed by 2T-physics. Now that 2T-physics has produced the $4+2$ version of this theory, the powerful mathematical techniques can be used to learn more deeply about the properties of the $N = 4$ theory itself as well as the properties of the extra dimensions, by studying not only the conformal shadow but also the other shadows that share the same powerful symmetries. These are projects for the future.

As already emphasized several times in previous sections, one of the main novelties in 2T-physics is that it produces many 1T-physics shadows from the same parent theory. Some of the remarkable properties of 2T-physics at the particle level that are summarized in Fig. 7.7 include surprising properties of the shadows, such as

- emergent spacetimes,
- emergent parameters,
- holography,
- hidden symmetries,
- unification of 1T dynamics by dualities.

These features of 2T-physics, which were not recognized before in the context of 1T-field theory, can be used to develop new mathematical techniques in field theory:

2T-physics provides new technical tools for computations in 1T-physics.

In particular, the hidden symmetries and dualities are the main tools that could be used for simplifying computations.

Dualities transform shadows with different $3+1$ geometries into one another. Shadows emerge from the different embeddings of $3+1$ dimensions in $4+2$ dimensions. These embeddings contain some parameters called " moduli" that appear in $3+1$ dimensions as parameters of the 1T shadow theory. Examples of such parameters include mass, curvature, or interaction with backgrounds, as indicated in Fig. 7.7. These parameters capture properties of different perspectives of the higher dimensions.

These phenomena exist also at the field theory level. Field theories related by duality transformations are considered to be the same theory under a different disguise. As seen in Fig. 7.7, the conformal shadow that connects 2T-field theory and the familiar 1T-field theory, such as the standard model and general relativity, is only one of many possible shadows. The other shadows represent different looking 1T-field theories, so these are the duals of the standard model and general relativity.

The shadows and dualities are most easily understood in the worldline formulation of 2T-physics. While the investigation of dualities in the 1T field theory formalism is ongoing, some of the simpler cases have been reported in [41][1] for mathematical details of the shadows in Fig. 7.7. for scalar fields and in [42] for Dirac and Yang–Mills fields. One of the important goals in the ongoing projects in 2T-physics is to make use of dualities to develop new computational tools to solve difficult mathematical and physical problems that have emerged as part of our current cherished theories. It is expected that such methods can eventually be usefully applied to non-perturbative studies of field theory, including quantum chromo-dynamics (QCD) which still awaits solution in the strongly interacting regimes.

Besides being a computational tool, dualities provide also information about the higher space–time. Since the shadows give different perspectives of the $4+2$ theory as viewed by observers that are stuck in $3+1$ dimensions, the information obtained from the dual theory could then be interpreted as observations of various features of the higher dimensional space–time. This is how observers in $3+1$ dimensions get to experience indirectly the effects of $4+2$ dimensions.

Through the dualities, and through hidden symmetries related to the higher space–time, the parent theory in d+2 dimensions provides a new kind of unification of various 1T-physics field theories.

Progress in 2T-field theory has not reached supergravity yet although this is close. It would appear that having constructed 2T-gravity as well as 2T supersymmetric field theory for Klein–Gordon, Dirac and Yang–Mills fields already provide the necessary ingredients to construct supergravity. However, there are additional technical steps that so far have not been completed to combine these building blocks together. When constructed, the 2T version of supergravity in 11+2 dimensions would yield the 10+1-dimensional supergravity as one of its shadows.

[1] See Tables I, II, III in [41] for mathematical details of the shadows in Fig. 7.7

How about 2T string theory? A first attempt to construct strings and branes in 2T-physics was discussed in 1999 [21]. It was reported that tensionless strings and branes in 1T-physics emerged naturally from a simple version of strings and branes, but that the tensionful string and branes did not. In a later development in 2004, the Witten–Berkovits twistor superstring was constructed as a shadow of its counterpart as a 2T twistor superstring in $4+2$ dimensions [32, 33]. However the usual string theory so far has resisted to be represented as a shadow.

> 2T supergravity, 2T strings, 2T branes, 2T M-theory only partially constructed in 2T-physics.

The main difficulty in obtaining the familiar tensionful 1T strings as shadows of a 2T string theory might be overcome by including the dilaton in the construction of the 2T string theory. That the dilaton is an important ingredient, which was missed in previous attempts, has now become apparent with the recent construction of 2T-gravity in 2008. So, among the challenging projects in 2T-physics is to find the general formulation of 2T string theory that would include shadows of tensionful strings.

Although the mysterious M-theory has not been constructed even in 1T-physics, it is known that its several known corners are connected to each other by dualities as illustrated in Fig. 10.1. This diagram is reminiscent of the 2T-physics shadows in Fig. 7.7, where dualities are intuitively expected for shadows that result from the same object. Similarly, it may be that the dualities of M-theory also arise in the same way. This will not be settled until M-theory itself is constructed, but the natural dualities that arise in 2T-physics is an indication that the 2T approach is a promising path to construct M-theory and explain its dualities.

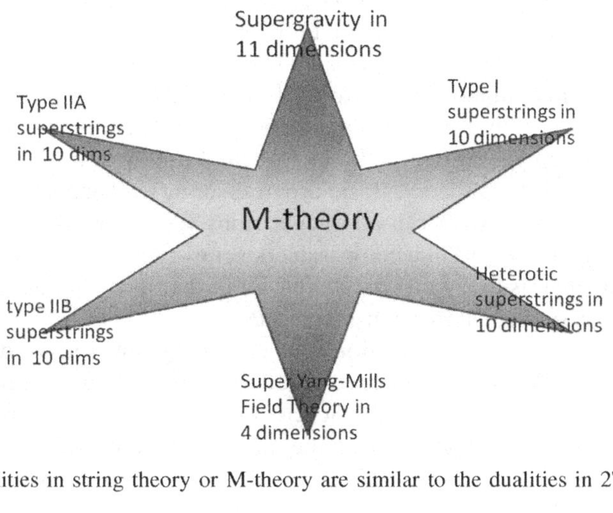

Fig. 10.1 Dualities in string theory or M-theory are similar to the dualities in 2T-physics as in Fig. 7.7

According to 2T-physics we should expect that one of the corners of 2T M-theory at "low energies" should be 2T supergravity in 13 dimensions, whose shadow is 11-dimensional supergravity shown as a corner of M-theory in Fig. 10.1. This suggests that 2T supergravity as well as 2T M-theory should obey supersymmetry in 13 dimensions. If constructed, this theory would be in 11+2 dimensions, whose global supersymmetry can only be the supergroup OSp(1|64). This is because OSp(1|64) is the smallest unique supergroup that contains the Lorentz symmetry SO(11, 2). This unique mathematical structure was already explored to some extent as the basic property of S-theory in 1996 [5–9]. The purpose of S-theory was to provide a unification of the symmetries of M-theory consistent with its dualities. Indeed OSp(1|64) has sub-symmetries consistent with 11 dimensions or 10 dimensions that are precisely type-IIA, type-IIB, heterotic, type-I supersymmetries, which are consistent with the dualities of M-theory in Fig. 10.1. It is encouraging that the unique supersymmetry OSp(1|64) demanded by a dynamical 2T-physics theory in 13 dimensions has those very attractive and desirable features explored through S-theory. This is an indication that the 2T approach is quite promising for the eventual construction of M-theory.

So far we have managed to develop 2T-physics fully at the single particle level and at the field theory level. Although the 2T-field theory formulation works well to fit nature, it leaves us with the feeling that perhaps there is a deeper field theory formulation:

A deeper phase space formulation of field theory is likely to exist.

This feeling comes by reminding ourselves that the fundamental concept behind 2T-physics is the momentum–position symmetry based on Sp(2, R) as illustrated in Fig. 7.3. The 2T-field-theoretic approach is able to incorporate the Sp(2, R) constraints, but in a formalism that favors position space over momentum space. So Sp(2, R) is taken into account in 2T-field theory in a form that blurs the position–momentum symmetry. This is not as symmetric between X^M and P_M as it was in the worldline formulation of 2T-physics. There should be a more fundamental field-theoretic approach with a more manifest position–momentum symmetry, perhaps with fields that depend on both X^M and P_M. In that case it would be likely to be a non–commutative field theory. Basic progress along this line that included fields as functions of phase space with all integer spins was reported in [26, 27]. If this avenue could be developed to a comparable level of success as the current 2T-field theory formalism, it is likely that it will go a lot farther than our current approach in predicting surprising phenomena in 1T-physics.

There are experimental observations that push the boundaries of our established theoretical ideas. We have mentioned a few of them earlier in this book. These include dark energy that causes the acceleration of the universe as in Fig. 10.2, the invisible dark matter that provides the missing gravitational pull within and in between the galaxies as in Fig. 10.3, the actual origin of mass which is related to the still obscure Higgs particle as in Fig. 9.4, the precise masses of the fundamental

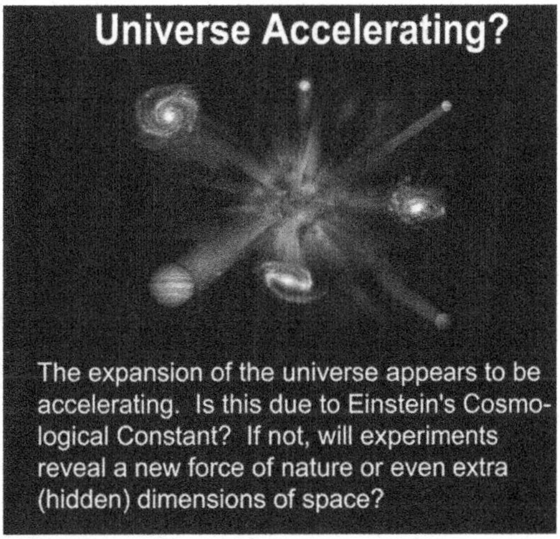

Fig. 10.2 Dark energy could be explained by the cosmological constant (Credits, Particle Data Group.)

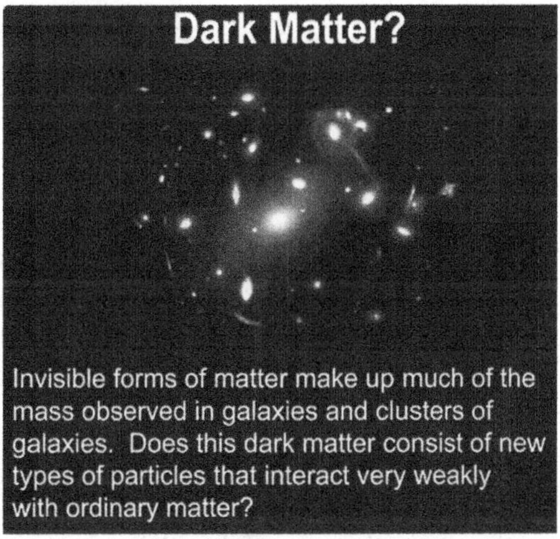

Fig. 10.3 Dark matter = weakly interacting massive particles (WIMPs)? (Credits, Particle Data Group.)

particles as in Figs. 9.5 and 9.6, neutrino mixing, origin of matter–anti-matter asymmetry in our universe as in Fig. 2.8, the strong CP violation problem of QCD and its missing axion.

There is no shortage of theoretical ideas in the context of supersymmetry, string theory, 2T-physics, and others that potentially can provide answers to each puzzle. However, at the present time the theoretical physics community is not confident about what really lies beyond the standard model and general relativity. At this stage we have many puzzles and many possible answers, which will be sorted out over time, either with the help of experiments or with the further development of theoretical concepts and computational techniques. It could be dangerous to stray too

Fig. 10.4 What's around the corner? (Credits, Brian Zaikowski.)

much off the established track as in Fig. 10.4, but one must still keep on searching by making some hypotheses that are consistent with the established knowledge.

Although incomplete, we are sure that the standard model and general relativity are part of the truth about our universe within the currently explored energy and space–time boundaries. So, they provide a boundary condition for testing the correctness of any theories that try to go beyond our firmly held ideas at the present. If a newly proposed theory turns out to be inconsistent with the standard model and general relativity, then it is bound to be wrong unless the disagreement is subtle and experimentally not ruled out. So we have some checks to correct our course when we wander too much off the established truths.

To guide our search of new truths it is important to emphasize that the phenomenologically successful field theories now have counterparts in $4+2$ dimensions in the form of 2T-physics that already provides an experimentally correct extension of the known formulation of the standard model and general relativity. It provides new perspectives on the significance of space and time and paves a new path for the unification of 1T-physics theories via the shadows of the 2T space–time.

In particular we have learned that

> The extra 1 space and 1 time dimensions in 2T-physics are neither small nor hidden.

This is in contrast to the Kaluza–Klein-type notion of tiny extra space dimensions, as well as in contrast to the "large" dimensions in brane-world theories discussed by Prof. Terning in the second part of this book. These two extra 1+1 dimensions are above and beyond the Kaluza–Klein or brane-world extra *space* dimensions. Although we are observers stuck in $3+1$ dimensions, without the privilege of traveling in $4+2$ dimensions, the extra 1+1 dimensions in 2T-physics can still be experienced indirectly by $3+1$-dimensional observers through the properties of the many shadows. The predicted effects, such as hidden symmetries and dualities, that are verifiable in $3+1$ dimensions provide the means for observing and interpreting the higher space–time. Does this kind of indirect observation mean that the extra dimensions actually exist? This can be debated at the philosophical level, but if the

Fig 10.5 New concepts are often resisted, but they cannot be ignored when they explain previously unknown parts of nature (Credits, ScienceCartoonsPlus.com.)

only obvious explanation of the correlations, dualities, and hidden symmetries of different shadows is the extra dimensions, then why even try to call it something else?

So 2T-physics provides new guidance for addressing the currently unsolved mysteries of nature, but is it needed? Yes it is needed because, as I described in this book, we are already familiar with phenomena that cannot be captured in the usual formalism and are systematically missed in 1T-physics. The more encompassing approach of 2T-physics should be pursued vigorously to make progress beyond our current understanding of nature.

As in the cartoon of Fig. 10.5, 2T-physics represents progress that cannot be ignored. It is very difficult for people to digest an entirely new view of space–time and the accompanying shift in the formulation of the basic equations of physics. I received only encouragement for and no objections to 2T-physics from my distinguished physicist colleagues. Yet their own contributions along this path have been slow. The excitement and preoccupation with one's own ideas create great resistance to change research direction. Furthermore, many young physicists feel unsafe in pursuing new paths of research and seek the assurance of the crowds in the choice of research problems. This combination creates a natural tendency to ignore new ideas until the pressure rises when many others jump on the bandwagon. This is a familiar pattern that has occurred all too often in the history of physics, including in the history of the currently popular string theory. 2T-physics should influence our thinking as we proceed from our current perspective, not only in search for the fundamental theory but also for reaping the benefits of the higher 2T-physics perspective in developing both physical insight and computational techniques in all aspects of 1T-physics.

There is a lot more to be done in 2T-physics. Future projects will deal with what I think will be the most powerful form of 2T-physics as a non-commutative 2T-field theory in phase space along the lines of [27]. The current research activities are concentrated on the study of 2T-physics at its foundational, phenomenological, and philosophical levels. Foundational aspects include developing field theory computational techniques for the 2T standard model and 2T gravity consistent with 2T gauge symmetries at the quantum level, and completing the formulation of 2T string theory to seek an eventual unification higher than envisaged before. Phenomenological aspects include developing experimental tests that could expose properties of the underlying higher space–time. Philosophical aspects include the interpretation of the shadows to establish the reality and meaning of the $4+2$-, or more generally d+2-, dimensional space–time that produce them.

Chapter 11
Further Reading

The following list of papers that appear in historical order contain the technical details of my effort, together with my collaborators', to understand extra time dimensions, leading to the construction of 2T-physics and its evolution to the current status. References to many but not all of these papers appear in the text.

The last four references are reviews based on lectures at various stages of this development.

References

1. I. Bars and S. Yankielowicz, "U-duality multiplets and non-perturbative string states", Phys. Rev. **D53** (1996) 4489, [arXiv:hep-th/9511098].
2. I. Bars, "Consistency between 11D and U duality", Nucl. Phys. Proc. Suppl. **49** (1996) 115, [arXiv:hep-th/9601164].
3. I. Bars, "Duality and hidden dimensions," Lecture in 1995 conference, proc. *Frontiers in quantum field theory*, Ed. H. Itoyonaka, M. Kaku. H. Kunitomo, M. Ninomiya, H. Shirokura, Singapore, World Scientific, 1996, page 52. [arXiv:hep-th/9604200].
4. I. Bars, "Supersymmetry, p-brane duality and hidden space and time dimensions," Phys. Rev. **D54** (1996) 5203, [arXiv:hep-th/9604139].
5. I. Bars, "S-theory", Phys. Rev. **D55** (1997) 2373, [arXiv:hep-th/9607112].
6. I. Bars, "Algebraic structures in S-theory", [arXiv:hep-th/9608061], appeared in the Proc. of 2nd International Sakharov Conference on Physics, and in proceedings of Strings-96.
7. I. Bars, "Black hole entropy reveals a 12th dimension", Phys. Rev. **D55** (1997) 3633, [arXiv:hep-th/9610074].
8. I. Bars, "A case for 14 dimensions", Phys. Lett. **B403** (1997) 257, [arXiv:hep-th/9704054].
9. I. Bars, "Duality covariant type IIB supersymmetry and non-perturbative consequences", Phys. Rev. **D56** (1997) 7954, [arXiv:hep-th/9706185].
10. I. Bars and C. Kounnas, "A new supersymmetry", hep-th/9612119.
11. I. Bars and C. Kounnas, "Theories with two times", Phys. Lett. **B402** (1997) 25, [arXiv: hep-th/9703060].
12. I. Bars and C. Kounnas, "String and particle with two times", Phys. Rev. **D56** (1997) 3664, [arXiv: hep-th/9705205].
13. I. Bars and C. Deliduman, "Superstrings with new supersymmetry in (9, 2) and (10, 2) dimensions," Phys. Rev. D56 (1997) 6579, [arXiv:hep-th/9707215];
14. I. Bars and C. Deliduman, "Gauge principles for multi-superparticles," Phys. Lett. B417 (1998)24, [arXiv:hep-th/9710066].

I. Bars, J. Terning, Founding Editor: F. Nekoogar, *Extra Dimensions in Space and Time*, Multiversal Journeys, DOI 10.1007/978-0-387-77638-5_11,

15. I. Bars, C. Deliduman and O. Andreev, "Gauged duality, conformal symmetry and spacetime with two times", Phys. Rev. **D58** (1998) 066004 [arXiv:hep-th/9803188].
16. I. Bars, "Conformal symmetry and duality between free particle, H-atom and harmonic oscillator", Phys. Rev. **D58**, 066006 (1998) [arXiv:hep-th/9804028].
17. I. Bars and C. Deliduman, "Gauge symmetry in phase space with spin, a basis for conformal symmetry and duality among many interactions", Phys. Rev. **D58** (1998) 106004, [arXiv:hep-th/9806085].
18. I. Bars, "Hidden symmetries, $AdS_d \times S^n$, and the lifting of one-time physics to two-time physics", Phys. Rev. **D59** (1999) 045019 [arXiv:hep-th/9810025].
19. I. Bars, C. Deliduman and D. Minic, "Supersymmetric two-time physics", Phys. Rev. **D59** (1999) 125004, [arXiv:hep-th/9812161].
20. I. Bars, C. Deliduman and D. Minic, "Lifting M-theory to two-time physics", Phys. Lett. **B457** (1999) 275, [arXiv:hep-th/9904063].
21. I. Bars, C. and Deliduman, D. Minic, strings, branes and two-time Physics, Phys. Lett. **B466** (1999) 135, [arXiv:hep-th/9906223].
22. S. Vongehr, "Examples of black holes in two-time physics," [arXiv:hep-th/9907077].
23. I. Bars, "Two time physics with gravitational and gauge field backgrounds", Phys. Rev. **D62**, 085015 (2000) [arXiv:hep-th/0002140];
24. I. Bars and C. Deliduman, "High spin gauge fields and two time physics", Phys. Rev. **D64**, 045004 (2001) [arXiv:hep-th/0103042].
25. I. Bars, "Two-time physics in field theory", Phys. Rev. **D 62**, 046007 (2000), [arXiv:hep-th/0003100].
26. I. Bars and S. J. Rey, "Noncommutative Sp(2,R) gauge theories: A field theory approach to two-time physics," Phys. Rev. D **64**, 046005 (2001) [arXiv:hep-th/0104135].
27. I. Bars, "U*(1, 1) noncommutative gauge theory as the foundation of 2T-physics in field theory," Phys. Rev. D **64**, 126001 (2001) [arXiv:hep-th/0106013].
28. I. Bars, "A mysterious zero in $AdS_5 \times S^5$ supergravity," Phys. Rev. D **66**, 105023 (2002), [arXiv:hep-th/0205194].
29. I. Bars, "Hidden 12-dimensional structures in $AdS_5 \times S^5$ and $M^4 \times R^6$ supergravities," Phys. Rev. D **66**, 105024 (2002), [arXiv:hep-th/0208012].
30. I. Bars, C. Deliduman, A. Pasqua and B. Zumino "A hidden twelve-dimensional superPoincare symmetry in eleven dimensions", Phys. Rev. **D69** 106007 (2004), [arXiv:hep-th/0312147].
31. I. Bars, "2T physics formulation of superconformal dynamics relating to twistors and supertwistors," Phys. Lett. B **483**, 248 (2000) [arXiv:hep-th/0004090].
32. I. Bars, "Twistor superstring in 2T-physics," Phys. Rev. **D70** (2004) 104022, [arXiv:hep-th/0407239].
33. I. Bars, "Twistors and 2T-physics," AIP Conf. Proc. **767** (2005) 3, [arXiv:hep-th/0502065].
34. I. Bars and M. Picón, "Single twistor description of massless, massive, AdS, and other interacting particles," Phys. Rev. **D73** (2006) 064002 [arXiv:hep-th/0512091].
35. I. Bars and M. Picón, "Twistor transform in d dimensions and a unifying role for twistors," Phys. Rev. **D73** (2006) 064033, [arXiv:hep-th/0512348].
36. I. Bars and B. Orcal, "Generalized twistor transform and dualities, with a new description of particles with spin, beyond free and massless", [arXiv:0704.0296 [hep-th]].
37. I. Bars and Y-C. Kuo, "Interacting two-time physics field theory with a BRST gauge invariant action", Phys. Rev. **D74** (2006) 085020, [harXiv:ep-th/0605267].
38. I. Bars, "The standard model of particles and forces in the framework of 2T-physics", Phys. Rev. **D74** (2006) 085019 [arXiv:hep-th/0606045].
(For a summary see: "The standard model as a 2T-physics theory", [arXiv:hep-th/0610187], which appeared in *SUSY06*, Ed. J. Feng, AIP Conf. Proc. Vol. 903 (2007), p. 550).
39. I. Bars and Y.C. Kuo, "Field theory in 2T-physics with $N=1$ supersymmetry" Phys. Rev. Lett. **99** (2007) 41801 [arXiv:hep-th/0703002].
40. I. Bars and Y.C. Kuo, "Supersymmetric field theory in 2T-physics," Phys. Rev. **D76** (2007) 105028,. [arXiv:hep-th/0703002].

41. I. Bars, S. H. Chen and G. Quélin, "Dual field theories in (d-1)+1 emergent spacetimes from a unifying field theory in d+2 spacetime," Phys. Rev. **D76** (2007) 065016. [arXiv:0705.2834 (hep-th)].
42. I. Bars, and G. Quélin, "Dualities among 1T-field theories with spin, emerging from a unifying 2T-field theory", Phys. Rev. **D77** (2008) 125019 [arXiv:0802.1947 (hep-th)].
43. I. Bars, "Gravity in 2T-physics", Phys. Rev. **D77** (2008) 125027 [arXiv:0804.1585 (hep-th)].
44. I. Bars and Y.C. Kuo, "N=2 and N=4 supersymmetric field theory in 2T-physics", Phys. Rev. D79 (2009) 025001 [arXiv:0808.0537 (hep-th)].
45. I. Bars and S. H. Chen, "Geometry and symmetry structures in 2T gravity ", Phys. Rev. D79 (2009) 085021, [arXiv:0811.2510 (hep-th)].
46. I. Bars and Y.C. Kuo, "Super yang mills theory in 10+2 dimensions, another step toward M-theory", in preparation.
47. I. Bars and S. H. Chen, "Is there a formulation of physics with more times beyond 2T-physics?", in preparation.

Reviews of 2T-physics at various stages of development appear in the following lectures:

48. I. Bars, "Two-time physics", in the Proc. of the 22nd Intl. Colloq. on Group Theoretical Methods in Physics, Eds. S. Corney at. al., World Scientific 1999, [arXiv:hep-th/9809034];
49. I. Bars, "Survey of two-time physics," Class. Quant. Grav. **18**, 3113 (2001) [arXiv:hep-th/0008164].
50. I. Bars, "2T-physics 2001," AIP Conf. Proc. **589** (2001), pp. 118–30; and AIP Conf. Proc. **607** (2001), pp. 17–29 [arXiv:hep-th/0106021].
51. I. Bars, "Lectures on twistors," (5 lectures), in Quantum Theory and Symmetries IV, Ed. V.K. Dobrev, Heron Press (2006), Vol. 2, p. 487 (Bulgarian Journal of Physics supplement, Vol. 33); and in Recent Trends in string/M-theory, Ed. J.X. Lu. [arXiv:hep-th/0601091].

Part II
Extra Dimensions of Space

John Terning

Chapter 12
The Popular View of Extra Dimensions

The possible existence of extra dimensions of space is one of the few topics of fundamental physics research that is deeply embedded in popular culture. When he was 4 years old, my son found a comic in a bubble gum wrapper that showed a boy wearing two sets of 3-dimensional glasses: the boy was trying to find the sixth dimension. So even 4-year olds know about extra dimensions. There is currently a TV show where characters routinely enter another dimension by stepping through "dimensional rifts." People of my age will remember a popular singing group from the 1960s called "The Fifth Dimension." A very popular TV show from the 1950s, "The Twilight Zone," started with the words "You unlock this door with a key of imagination, beyond it is another dimension. . ."

Popular interest in extra dimensions stretches all the way back to Edwin Abbott Abbott's 1884 novel "*Flatland: A Romance in Many Dimensions*" [1] wherein the central character, a square who lives in a 2-dimensional world called flatland, encounters a sphere from spaceland. The sphere tries to explain that he exists not only in the 2 dimensions of flatland but also in a third dimension which the square cannot perceive called "up." The sphere is unable to explain using mere words and logic what he means by "up" and in frustration lifts the square out of flatland. The square's world view is shattered; he then has an epiphany and begs to be shown the fourth and fifth dimensions as well, but the sphere considers this proposal to be impossibly ridiculous and drops the square back into flatland. To his fellow flatlanders the square seems to miraculously reappear out of nowhere. The unfortunate square cannot explain his adventure to them but only say that he was moving through an extra dimension called "up" that he cannot point to since it is perpendicular to all the dimensions they can see. Considered a dangerous lunatic by the flatlanders the square ends up in prison unable to come to grips with his journey through the extra dimension "up."

If extra dimensions really existed many strange events, like disappearing into and reappearing somewhere else from an extra dimension, might be possible and it is probably partly these bizarre possibilities that account for much of the interest in extra dimensions. Two other contributing factors are that Einstein showed us that time is really a fourth dimension and that string theory seems to require 11 dimensions. These last two ideas appeal to the seemingly deep human interest in

I. Bars, J. Terning, Founding Editor: F. Nekoogar, *Extra Dimensions in Space and Time*, Multiversal Journeys, DOI 10.1007/978-0-387-77638-5_12,

Fig. 12.1 Title page graphics drawn by Edwin Abbott Abbott himself

finding a fundamentally deeper level of reality than what is apparent in everyday life.

12.1 Einstein and the Fourth Dimension

After Einstein explained how time should be considered as the fourth dimension and his prediction of the bending of light was observationally confirmed during a total solar eclipse he became what we would now call an international superstar. In 1931 Charlie Chaplin invited Einstein to be his guest of honor at the Los Angeles premier of Chaplin's new silent film, "City Lights." When Einstein entered the theater filled with Hollywood actors, directors, and producers he received a standing ovation! Later Chaplin explained to Einstein [2], "They cheer me because they all understand me, and they cheer you because no one understands you."

Einstein's first claim to fame was his theory of special relativity, which required that we think of time as a fourth dimension, and that we modify our usual notions of space and time as separate quantities. Now there is a trivial sense in which time can be thought of as a fourth dimension. If I ask you to meet me in San Francisco at the corner of Clay and Montgomery, you would go to Google and easily find this location. The intersection of Clay and Montgomery is enough to specify a location on the 2-dimensional surface of the Earth. Arriving at that location you would quickly realize that you need a third piece of information: the altitude. To specify this third dimension I could tell you that we'll meet on the 43rd floor. But there is still one crucial piece of information missing! I need to tell you the time of our meeting, September 25 at 2:45 PM, otherwise we will miss each other even though we both go to the same point in space. So time is a fourth dimension, but there are other numbers I could tell you as well, the temperature, for example, or the pollen count. When we talk about a fourth dimension we mean much more than just a list of four numbers. There is a much more profound sense in which time is the fourth dimension, and to understand this we need to go back to James Clerk Maxwell's studies of electric and magnetic fields in the early 1860s.

By synthesizing the results of experiments by many other scientists on electricity and magnetism Maxwell was able to summarize everything that was known about these two subjects in four equations. These four equations, now called the Maxwell equations, provided an amazingly deep view of how our world works. First of all they showed that electricity and magnetism are just two different aspects of what has come to be known as the electromagnetic field. Electric and magnetic fields are now the stuff of everyday experience. Electric fields move electrons through our light bulbs and computer chips. The magnet in a compass aligns itself with the magnetic field of the Earth and tells us which way is north. The magnetic fields of moving magnets are used in generators to produce electric fields while the magnetic fields of stationary magnets hold shopping lists and kindergarten art onto refrigerators. Anyone who has fiddled with a magnet has a basic grasp of how fields work. The magnet acts as a source for the magnetic field and the field can reach out across space to produce a force on a piece of iron or another magnet. As two magnets are brought closer together the force gets stronger. If you have had a chance to play with a magnet and iron filings then you have also seen that the filings act like miniature magnets and tend to line up in a definite pattern of lines radiating from the north pole of the magnet to the south pole of the magnet as shown in Fig. 12.4. These field lines show not only the direction that the individual iron filings line up but also the strength of the field. Near the poles of the magnet the field lines are converging so the lines are denser: this is where the field is strongest. Far from the poles of the magnet the field lines are spread out: here the field is weaker.

By analyzing his equations Maxwell deduced that they had a very interesting type of solution: waves of electric and magnetic fields. These electromagnetic waves are ripples in the strength of the electromagnetic field, just as water waves are ripples in the height of water and sound waves are ripples in air pressure. Electromagnetic waves can be sent out like ripples in a pond, but they can travel equally well in any of the three directions of space. Just like ripples in a pond they travel at a definite speed, but the speed of electromagnetic waves is much greater than the speed of a water ripple. When Maxwell put in the experimentally determined values he needed to calculate the speed of electromagnetic waves he found it was very close to the speed of light, 186,000 miles/s (or 300,000 k/s). Maxwell then realized that what we perceive as light is just a type of electromagnetic wave. The German scientist Heinrich Hertz later showed that Maxwell was indeed correct by producing

Fig. 12.2 Charlie Chaplin and Albert Einstein, 1931

Fig. 12.3 James Clerk Maxwell, engraving by G.J. Stodart from a photograph by Fergus of Greenock

electromagnetic waves in his laboratory. Hertz's demonstration quickly led to the development of wireless telegraphy, radio communications, and ultimately wireless Internet connections.

Maxwell was also interested in how our eyes respond to light, while our ears can respond to a wide range of frequencies of sound (we perceive different frequencies

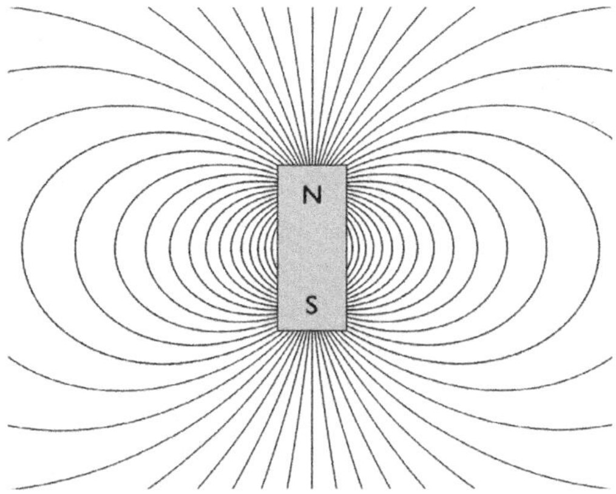

Fig. 12.4 Magnetic field lines around a bar magnet. Iron filings align themselves with the magnetic field lines

as different pitches) from about 20 to 20,000 oscillations per second (one oscillation per second is also known as 1 hertz, named in honor of Heinrich). On the other hand, or organ in this case, the retinas of our eyes respond only to electromagnetic frequencies between 400 THz (1 terahertz is a million, million hertz aka 1,000,000,000,000 Hz) and 790 THz, just a tiny sliver of the possible frequencies. It is quite remarkable that our ears cover frequencies that range over a factor of a thousand, while our eyes respond to frequencies that range over only a factor of 2. This restricted range is partly because of the types of light receptors in our eyes. Maxwell had his friends and acquaintances compare different colors with mixtures of red, green, and blue. His experiments helped to establish that our eyes have three types of color receptors which each respond to mostly either red, green, or blue [3]. This is the reason that TV screens and LCD displays only have three types of pixels which produce just red, green, or blue.

It is a somewhat curious fact that the narrow range of frequencies that we can see with our eyes falls inside the narrow range of frequencies that are not absorbed in water. It is less curious when we realize that the light receptors in the rods and cones of our retinas evolved when all animals lived under water. Another seemingly curious fact is that insects (bees, for instance) can see ultraviolet light which has a frequency slightly higher than we can see. It turns out that insects use the same light-sensitive molecules in their eyes as we do, and insects split off from our family tree after vision first evolved. It is actually the lenses of our eyes that block ultraviolet light. Our lenses evolved later on in order to provide an accurate focusing system that can adjust to focus on objects at different distances. Our retinas can respond to ultraviolet light, as has been experienced by people who have had their lenses removed because of cataracts [4]. The most famous case was the postimpressionist painter Monet, whose paintings of water lilies changed after his cataract surgery. The paintings in this period are blurrier, due presumably to his inability to focus properly, and are bluer, which other cataract sufferers with removed lenses have reported is how ultraviolet light looks to them.

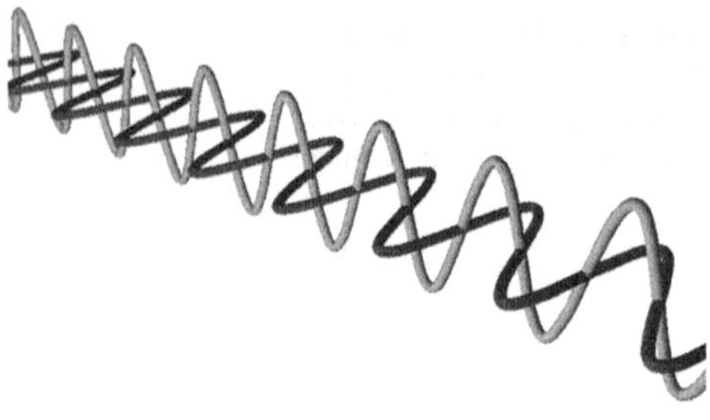

Fig. 12.5 An electromagnetic wave where the electric field is wiggling in the vertical direction and the magnetic field (shown in *dark*) is wiggling in the horizontal direction

If we precisely measured the electric field with a line of detectors as an electromagnetic wave passed by we would indeed find that the strength of the electric field wiggled up and down very much like a ripple in a pond (see Fig. 12.5). We would see the same thing if we measured the magnetic field as well. At first glance the biggest difference is that the wiggles (or *oscillations*) of the electromagnetic field can occur much more rapidly than those in a pond. Maxwell's equations implied one other important difference between the two types of waves. With a water wave a person standing in a boat and a person standing on the ground would measure different speeds for the wave. If the person on the boat was moving into the waves he would see the waves moving faster relative to him than what the person standing on the ground would record. The relative velocity is the sum of the individual velocities, just as a head-on collision of two cars each traveling at 60 miles/h toward the other is more dangerous than a 60 mile/h car colliding with a stopped car because the relative velocity is twice as big in the first collision as it is in the second. However, for a light wave, according to Maxwell's equations, the speed of the wave is always 186,000 miles/s, it doesn't depend on how fast the observer is moving, it doesn't even depend on how fast the source of the wave is moving. This seemed to be a great paradox to physicists at the end of the 19th century. If there is a wave, they reasoned, there must be some substance that "waves." For a water wave it is obviously the surface of the water that oscillates, but for a light wave it is not so obvious what the substance is that carries the oscillations. The physicists of the time decided to call this mysterious substance the "aether." But if such a substance really fills all of space then it is clear that as the Earth moves around the Sun, sometimes we should be moving along the flow of the aether and sometimes against the flow. Then if we measure the speed of light very carefully along different directions we will see the speed of light changes, just like the water waves.

People tried to make more and more precise measurements of the speed of light in order to find this variation. Most famously Michelson and Morley in Cleveland made extremely precise measurements of the speed of light but failed to detect any variation due to the supposed aether. So, around the turn of the 20th century physicists believed that all of space was filled with aether, but no one could detect any sign of it. Efforts were continued to make even more precise measurements in the hopes of detecting some sign of the aether. Other physicists tried to fool around with Maxwell's equations in order to make the equations fit better with common sense. There was a third group of scientists, consisting of Albert Einstein, who said that if we assume that the equations are correct and light always travels at the same velocity, then we need to change our common sense notions of space and time.

Einstein liked to perform "thought experiments" ("gedankin experiments" in German) in order to clarify how the world works. Suppose that Anna is standing on a platform with her arms held out to either side and she sets off a flash of light from a bulb in each of her hands at the same time. Assuming her arms are of equal length the flash of light from the left will hit her nose at exactly the same time as the flash of light from the right does (see Fig. 12.6). Now suppose we put Anna on a high-speed train and have her repeat the experiment with one hand toward the front of the train and one hand toward the rear. Einstein pointed out that no experiment has shown a dependence on the speed of the laboratory. As long as the laboratory moves at a constant speed (that is there is no acceleration) its motion doesn't matter. Indeed experiments would be almost impossible to understand if they depended on velocity, since what would be the reference point that we measure the velocity from? A point on the surface of the Earth is revolving around the center of the Earth, the Earth moves around the Sun, the Sun and other stars in our galaxy move around the center of the galaxy, our galaxy moves around the center of our cluster of galaxies, etc. Anyone who has ridden a train might have had the experience of the train starting up so smoothly that one has the impression that the nearby trains and even the train station have started moving. So experiment and experience teach us that the laws of nature do not depend on how fast we move, if there is no acceleration involved, and we expect that when Anna repeats her experiment on a high-speed train that she will find exactly the same result no matter how fast the train is moving: the light from the left and the light from the right will meet at her nose at exactly the same time.

Now consider how the situation looks to Bob who is standing still beside the tracks and watching Anna rush by. He must agree that the light from left and right meet at Anna's nose, but the light from one side has to go farther than the light from

Fig. 12.6 Two lights are flashed on simultaneously and meet at Anna's nose in the middle sometime later

the other side. Bob sees one of Anna's hands ahead of her in the direction she is moving and one hand trailing behind her. When the light leaves the trailing hand, Anna is rushing away from it, while she is rushing into the light coming from the leading hand. In the amount of time it takes for the light from the leading hand to reach Anna's nose, the light from the trailing hand could not have completed its journey since during this time Anna has been moving away from the light. But we already know that the light from the trailing hand and the light from the leading hand have to meet at the same time at Anna's nose. The only way out of this conundrum is for the light from the trailing hand to leave first so that it has more time to complete the journey to Anna's nose. Figure 12.7 shows the sequence of events that Bob must see.

Fig. 12.7 The same two lights are flashed on but viewed by someone moving with respect to Anna. From this point of view the lights cannot turn on simultaneously if both are to meet at the nose at the same time

What Einstein has taught us is that two events (like the lights turning on) can be simultaneous for one person but happen at different times for someone else if these two people are moving with some relative velocity. This basic fact, that simultaneity depends on *relative* velocities, is the corner stone of Einstein's special theory of relativity. The other well-known surprises of special relativity (length contraction, time dilation, the speed of light being the maximum possible speed) can be derived from the relativity of simultaneity using simple logic and some high school algebra.

The relativity of simultaneity is what shows us that time is a fourth dimension like the three spatial dimensions that we all experience in our day-to-day lives.

To see why time is a dimension we need to think more deeply about what we mean by dimensions. Let's think about a simple situation with only 2 dimensions: making a map, using the directions east and north. We usually draw a vertical light to represent moving towards the north, so that places that are further north are further up the page and places that are further south are further down the page. Similarly we draw places that are further east to the right and further west to the left. Suppose you decide to put your new GPS unit to use by making your very own map. You center your map on your house, so that all locations are specified by how many miles east or west and/or north or south of your house they are. Your office is 3 miles east and 4 miles north, and placing this information on your map you find, by recalling a dim memory of the Pythagorean theorem, or by just measuring on your map, that your office is 5 miles away as the crow flies. So far, so good. Now your neighbor, Fred, insists on helping you out using his GPS unit that he got on e-Bay. However his GPS doesn't seem to be calibrated correctly since it insists that north is more like what you would call northwest. Undeterred, Fred gets to work and finds that by his reckoning your office is 4.9 miles east and 1 mile north of your house. However he still finds that it is 5 miles to your office as the crow flies. This is as it should be: the proverbial crow doesn't care about your GPS units, and the distance between your house and your office can't depend on what direction you decide to call north (north could point toward the North Pole, or the direction a compass points, which is "magnetic north"). But if we choose a different direction for north it does change how far north your office is. This is because when we choose a different direction for north, the new distance north is a mixture of the old distance north and the old distance east (see Fig. 12.8).

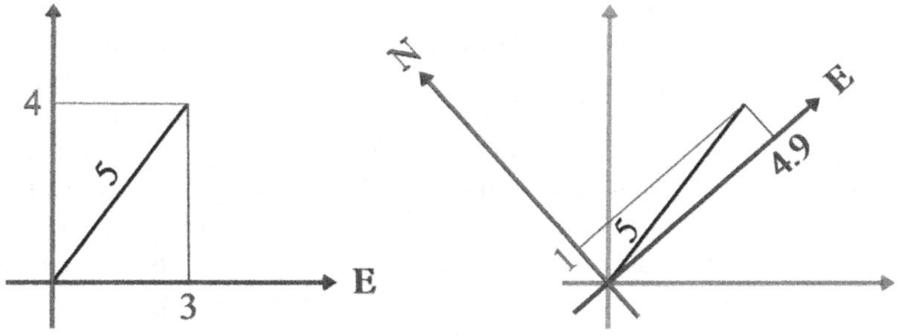

Fig. 12.8 The distance between two points does not depend on which direction we choose to call north, but the part that is north and the part that is east do depend on this choice

The arbitrariness of what to call north may seem like an annoying source of confusion, but to a physicist such arbitrariness is an opportunity to exploit. Such arbitrariness means that no matter which way we choose we will get the same answer, and so the fundamental properties involved can't depend on this choice. A physicist would say that the arbitrariness reflects an underlying *symmetry*. Now symmetry may seem like an odd choice of words, since to most people symmetry simply means that if you fold a picture down the middle the two halves match up, or equivalently holding a mirror up to the center line completes the whole picture. This type of symmetry is more precisely called a *reflection symmetry*; see Fig. 12.9.

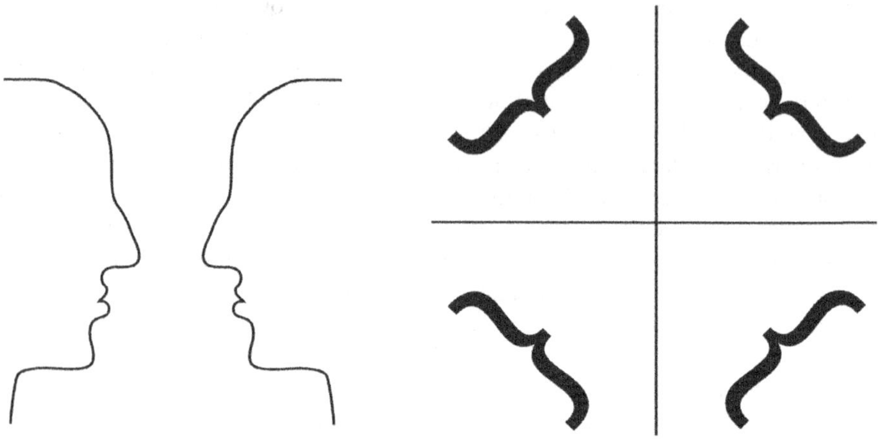

Fig. 12.9 The figure on the *left* has a reflection symmetry, holding a mirror up to a *vertical line* midway between the two profiles will reproduce the whole figure. Alternatively, folding the paper along the midway line will align one profile on top of the other. The figure on the *right* has two reflection symmetries, one around the *vertical line* and one around the *horizontal line*

There are many other types of symmetry in addition to reflection symmetry. The next simplest type of symmetry is *rotation symmetry*. Look at the top-left diagram in Fig. 12.10 which has three dots equally spaced around a circle. This diagram has a reflection symmetry around a vertical line running down the center. It also has a rotation symmetry, we can simply rotate the page around the center dot by 120° clockwise, which moves the dot on the left to the top position, and the dot at the top to the position of the dot on the right. The rotation has to be 120 degrees because we want to rotate the circle one-third of the way around, and 360° would rotate the circle all the way around, which just takes it back to its original position. The diagram on the top-right has four dots and reflection symmetries around both a vertical line and a horizontal line. It is also left the same by a quarter rotation (a 360/4=90° rotation). The diagram with five dots is invariant under a rotation of 360/5=72°. The diagram with six dots is invariant under a rotation of 360/6=60°. The diagram with eight dots invariant under a rotation of 360/5=45°. The bottom-right diagram with a ring can also be thought of as being the same as a diagram with an infinite number of dots. It is left the same by any arbitrarily small rotation. It is invariant under any rotation at all. In other words the diagram has a *continuous rotation symmetry* rather than just the discrete choices of rotations of our previous examples.

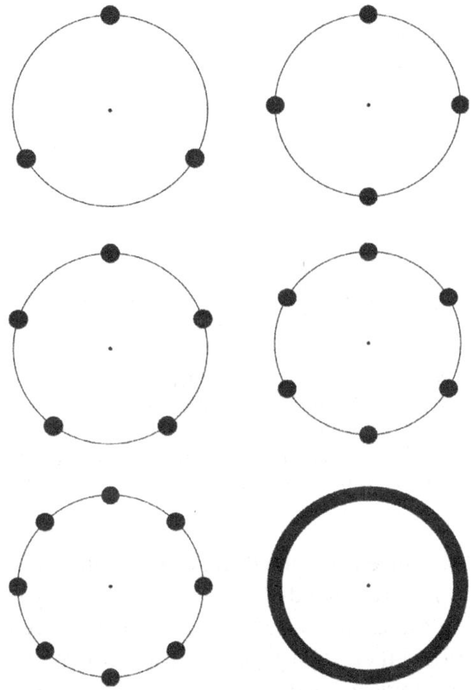

Fig. 12.10 The diagram with *three dots* around the *circle* has a reflection symmetry and a rotation symmetry for 120° rotations. The diagram with *four dots* has two reflection symmetries and is symmetric under 90° rotations. The diagram with *five* (*six*, *eight*) *dots* is symmetric under 72° (60°, 45°) rotations. The *bottom-right* diagram with a ring is symmetric under arbitrary rotations; it has a continuous rotation symmetry

Returning to our map, what we have in this example is a region (or space) with 2 dimensions, and we have arbitrarily chosen to call one direction north and a perpendicular direction east. Since there is a completely free choice of direction there is a *symmetry* that allows us to change this choice to any other choice. The symmetry here is just a rotation, as simple as a turn of the head. We can choose to rotate the direction we call north (which just amounts to rotating directions on our map). This rotation leaves the distances between two places the same: in other words the distance between two points is unchanging or *invariant* under rotations. But our coordinates (the distances east and north from your house, which is our reference point) do change when we rotate our choice of directions: the coordinates mix under the rotation. This simple example brings out two more key properties of dimensions. In addition to needing some number of dimensions to specify our position in space, the symmetries of the space can mix the coordinate values measured along these dimensions among themselves, and there is an invariant combination of the coordinate values that tells us how far apart two places are.

Now we are ready to see why time is a dimension. Let's make a map of space and time that shows us what is happening in Anna's experiment. We will draw the time direction vertically on the paper and take Anna's arms to point along the east–west direction (see the first part of Fig. 12.11).

By choosing the scales of each axis correctly we can have the light beam (moving at the speed of light, obviously) appear on our space–time map as a diagonal line.

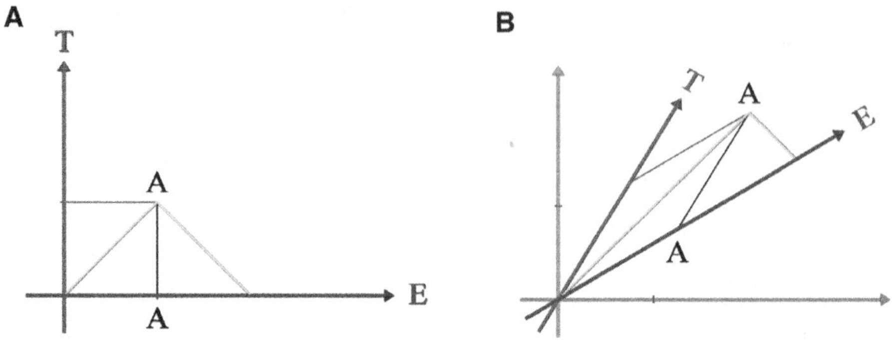

Fig. 12.11 A space–time "map" of the emission of the simultaneous light flashes described in the text; light rays travel along *diagonal lines*. The map on the *left* is from the point of view of someone moving with the light bulbs and the line from A to A represents Anna's nose (where the light flashes eventually meet) moving through time. The map on the *right* shows the point of view of someone like Bob who sees the light bulbs moving toward the east

The speed of something in the map can be inferred by taking the distance traveled (the horizontal distance) and dividing by the time taken (the vertical distance). If we measure time in seconds and the distance in light-seconds (the distance light travels in 1s) then light beams will be drawn on the map as diagonal lines. (Alternatively we could measure the distance in feet and time in "light-feet," the time it takes for light to travel one foot, and get a similar map.) An object at rest, like Anna, doesn't move, so it covers no distance in any amount of time and is just represented by a vertical line. A fixed time for Anna is represented by a horizontal line. In the first part of Fig. 12.11, Anna's position is represented by the vertical line in the center. When she turns on the lights in her hands the light moves along the diagonal lines, one from her left hand and one from her right, and the two light beams meet at her nose in the center.

The second part of Fig. 12.11 shows the space–time map from Bob's perspective. He also sees the two light beams meet at Anna's nose, but since Anna is moving to the east, the light from her trailing hand (to the west) has to have a head start and the light from her leading hand leaves later. Also since Anna is moving she is no longer represented by a vertical line, her line tilts a little toward the east. It is clear now that fixed positions for Anna are moving with respect to Bob, so what Anna would draw as a vertical line, Bob would draw as a line that is tilted away from vertical. Similarly, fixed times for Anna, like the moment she turns on the lights, are not simultaneous for Bob, so while Anna would represent that moment by a horizontal line, Bob draws it as a line that is tilted away from horizontal.

This is very similar to what happened in our example with the 2-dimensional map of the area around the house and office; however, here it is time and east that are being mixed together rather than north and east. Instead of choosing an arbitrary direction to be called north, here Anna has a particular direction she perceives to be time. But Bob has a different time direction since he is moving with respect to Anna. Bob's time direction is a mixture of Anna's time direction with a little bit of east thrown in. Neither choice is better than the other; we can make different choices by changing our velocity.

Also, just as rotations always leave the distances between two points untouched, or invariant, changing our velocity also leaves something invariant. It turns out that the invariant in this case is the square of the time separation subtracted from the square of the spatial separation. This sounds a little odd but it is more similar to rotations leaving the distance invariant than it sounds. Leaving the distance invariant is the same as leaving the sum of the squares of the distances in the two perpendicular directions invariant (we can thank Pythagoras of Samos for this insight). So what is really different about the invariant for velocity changes is that the square of the time separation contributes negatively while the squares of spatial separations contribute positively for both velocity changes in space–time and as well as for ordinary rotations in space.

If the preceding paragraph sounds too mathematical the summary is just that time is a dimension because changes in velocity mix time and space much as rotations can mix two spatial directions. Einstein revealed to us that time is a fourth dimension in a very deep sense. It took so long for humankind to recognize this fact because in everyday life we do not have much experience with velocity changes that come anywhere close to the speed of light. It was only by trying to understand light itself that we were led, by Einstein, to recognizing a fourth dimension.

12.2 Traditional Extra Dimensions

Given Einstein's revelation that time is a hidden fourth dimension it didn't take long for people to speculate about a hidden fifth dimension. The first concrete proposal for such a hidden spatial dimension came out of the work of Theodore Kaluza, later extended by Oskar Klein. The Kaluza–Klein theory, as it came to be known, supposes that the fifth dimension is hidden by being a very small circle rather than being infinite like the spatial dimensions that we already know.[1] The simplest analogy we can make is that of a thin tube. A microscopic being living on the tube would see a 2-dimensional surface, and if it was sufficiently small compared to the radius of the tube, the tube would seem to be flat, just as the surface of the Earth seems flat to us. An intelligent microscopic being would have no trouble recognizing a rotational symmetry in its world, so that it didn't matter which direction it labeled "north" (see Fig. 12.12.). But if this creature went on a long enough journey in the right direction it would find that it came back to its starting point, while going in the perpendicular direction this does not happen. A creature that was comparable in size to the thickness of the tube would of course immediately recognize that the two directions

[1] At least they seem to be infinite. We can only see 14 billion light-years out into the Universe. Since the Universe is about 14 billion years old this is as far as light could have traveled in the time since the Big Bang.

Fig. 12.12 Two directions on a tube look essentially the same close up

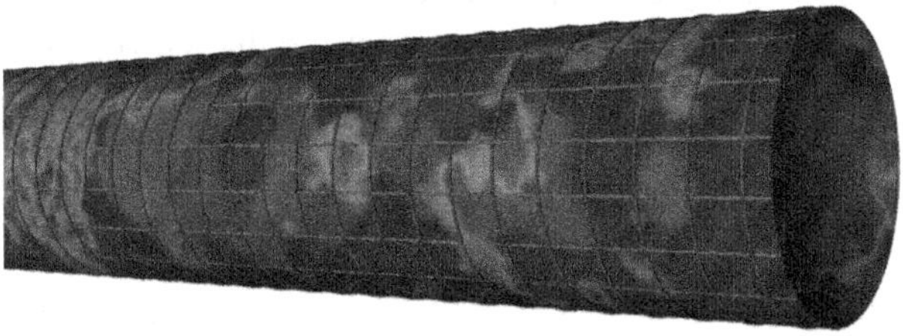

Fig. 12.13 The two directions on a tube look very different with a wider view where we can see that one could travel much farther in one direction than the other

were very different (see Fig. 12.13). A very large creature (with correspondingly less acute vision) might even believe the tube was 1 dimensional, just a line.

The question arises, if there is an extra dimension in our world that is smaller than the size of an atom, how would we be able to discover it? Another simple analogy that we can make is walking into a dark room full of stringed instruments. Even in the dark we can tell the difference between a violin, a guitar, or a cello by just tapping the strings. Even when they are playing the same note, these instruments sound different because of the overtones they produce. What we need to do in order to find extra dimensions is to bang on them and find the overtones.

To understand this a little better it would be helpful to know a little about waves. When we pluck a string on an instrument it can vibrate in many different ways. If we take a high-speed flash picture of the string as it vibrates we can see quite complex shapes depending on how we actually plucked the string. It turns out that any complicated vibration pattern can be thought of as being made out of simpler basic vibrations like those shown in Fig. 12.14 with a fixed number of regular wiggles. These basic vibrations are called standing waves, since they do not move through space like ordinary sound and light waves, they just sit in one place. When we pluck

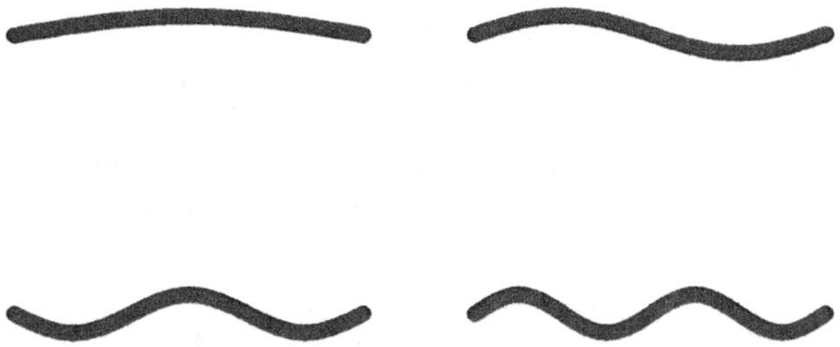

Fig. 12.14 A plucked string can vibrate in a variety of different ways, with different numbers of wiggles. The size of the vibration is, of course, exaggerated

a string near the center, the pattern we create usually is mostly composed of the first pattern with one wiggle, but there is also some contribution with three wiggles as well as other possibilities. The basic vibrations of all sound different to us because they oscillate at different rates, that is they have different frequencies. The string with two wiggles will jiggle (oscillate) twice as fast as the string with one wiggle. For example if our string with one wiggle in it jiggles up and down 440 times a second (that is, it has a frequency of 440 Hz, which we hear as a tone with the same pitch as middle A on a piano) then with two wiggles it jiggles 880 times a second (which is 12 notes up the keyboard, a high A).

With a musical instrument the vibrations of the strings also set up vibrations of the body of the instrument. These complicated vibrations can also be made out of simpler basic vibrations, although these simpler vibrations can themselves be quite complicated since they depend on the details of the shape of the instrument. If the frequency of the vibration of the body matches the frequency of the jiggling of the string then the body will vibrate in resonance with the string. If you have ever seen a singer shatter a glass with her voice, then you've seen how efficiently energy can be transferred when a frequency matches the resonant frequency of an object. This is just like pushing a swing, if you time your pushes correctly, a series of small pushes can make the swing go very high. In a musical instrument the string can get the whole instrument vibrating if the frequencies match. Since the body of the instrument has a much larger surface area it can transmit the vibrations to the air much more efficiently, and that is how we finally get to hear the tones produced. Now we see why different types of instruments sound different. The different body shapes have different resonant frequencies, and they enhance different amounts of the basic frequencies present in the jiggling of the string.

Now, turning back to extra dimensions if there is anything that can jiggle in the extra dimension then we can create standing waves in the extra dimension. For example if electromagnetic waves can move through the extra dimension then there should be standing electromagnetic waves in the extra dimension. You may not know it, but your microwave oven is a device that creates standing electromagnetic waves. These waves typically have a distance of about 5 in. (12 c) between the peaks of the wiggles. A general property of waves is that the frequency is equal to the speed of the wave divided by the distance between peaks. This is why, for a fixed length of string, more wiggles means a higher frequency. The more the wiggles, the shorter the distance between the peaks. So for a microwave the frequency is given by the speed of light divided by 12 c, which gives about 2450 million Hz. As we have seen, this is about 200,000 times smaller than the frequency of light, which just reflects the fact that the peaks of the wiggles of visible light are 200,000 times closer together than those of microwaves.

Fig. 12.15 Two possible excited Kaluza–Klein waves with three and five wiggles, respectively, wrapped around a circular extra dimension

So far we have been talking about light as if it is purely a wave, but, as the founders of quantum mechanics learned to their amazement, light comes in discrete packets of energy, or particles, that we call photons. When there are many photons traveling together with similar energies, then they behave exactly like a wave, but when we have just a single photon it behaves distinctly like a particle. Max Planck started the quantum revolution by studying what kind of light is emitted by a heated oven with a small window. Treating light as a wave gave good predictions for small frequencies, but it was only by assuming that light was emitted in discrete packets of energy that he could account for the high-frequency light. In fact Einstein received his Nobel prize not for the deep understanding of space and time provided by special relativity, but for his work on the photoelectric effect (which is how solar electric cells work). Einstein confirmed that photons must carry an energy proportional to their frequency, so that it takes a lot of energy to produce a photon of light with a high frequency. It turned out that this seemingly strange particle/wave duality applied not only to light but also to everything that we see in nature. While a large object, like a person, does not behave like a wave, if we examine a single electron in her body, the electron can behave like a wave.

A further surprise of the 20th Century was that combining special relativity with quantum mechanics requires the existence of anti-matter. Paul Dirac was the first to notice that incorporating the wave behavior of an electron with special relativity required an anti-electron (aka positron) with the same mass as the electron but the opposite (positive) charge. When an electron and a positron are brought together they can annihilate into pure energy, usually two photons. By the same token when two photons collide they can turn into an electron and a positron. Positrons were discovered shortly after Dirac published his results, and anti-particles that correspond to all the known particles have also been found. (The photon is a special case, it has no charge and it is its own anti-particle.) Today positrons are fairly commonplace, playing a key role in the PET scans[2] that are routinely used in medical diagnostics.

Let's return to thinking about a standing electromagnetic wave in an extra dimension one more time. If we assume that the extra dimension is in the form of a circle

[2]PET stands for positron emission tomography.

like Kaluza and Klein proposed, then the size of the circle will determine the distance between peaks of the wiggles of any particular wave. If there is one wiggle then the distance between peaks is just the circumference of the circle, since if we start at the peak and go all the way around the circle we get back to the one and only peak. If there are two wiggles, then the distance between peaks is half the circumference, and with three wiggles we have a third of the circumference. Each of these types of wiggles has a corresponding frequency, and the frequency grows with the number of wiggles. To create one of these wiggles we need to supply an energy proportional to the frequency, so we need more energy to create a wave with more wiggles. When we create ordinary particles energy is always conserved. For example, when an electron and a positron are produced from two photons the energy of the two photons adds up to the same value as total energy of the two electrons provided that we keep track of both the energy of the motion of the electrons (called the kinetic energy) and the mass energy. This is another consequence of special relativity: the energy equivalent to a certain amount of mass is given by Einstein's famous formula energy equals mass times the speed of light squared. For our example of producing an electron and a positron, this formula tells us the minimum amount of energy the photons must have. If we produce the electron and positron so that they are not moving (so that they have no energy of motion) then all the energy of each photon is converted into the mass of the electron or positron. Since photons can have arbitrarily small frequencies they can be produced with arbitrarily little energy, in other words photons are massless. For our standing waves in the extra dimension there is also a minimum energy required to produce each one: the energy proportional to the number of wiggles. This energy is not associated with motion in our ordinary dimensions, it is associated with the wiggles in the extra dimension. To us the energy behaves just like a mass.[3] So these standing electromagnetic waves in the extra dimension would behave like photons with a mass, and there wouldn't just be one new heavy photon, there would be a whole series with different masses that go up in mass proportional to the series 1,2,3, . . ., that is proportional to the number of wiggles in the extra dimension. So this is what experimentalists can look for in order to find an extra dimension: a series of particles regularly spaced in mass. In honor of Kaluza and Klein such a series of particles is called a tower of Kaluza–Klein particles. There is one more piece of the tower that I have left out of the story so far, that is the standing electromagnetic wave with no wiggles along the extra dimension, which can be created with an arbitrarily small energy, but this is just the ordinary photon. In general there would be such a Kaluza–Klein tower for every type of particle that can move through the extra dimension.

There is another question that you might be wondering about at this point: where is the extra dimension? Well, if it exists, then there is a little extra circle at every point in our ordinary space. If we produce a Kaluza–Klein particle with exactly the minimum energy required, then we can think of it as being produced on one particular circle, but if there is any extra energy the Kaluza–Klein particle will be

[3] Up to a factor of the speed of light squared, from Einstein's formula.

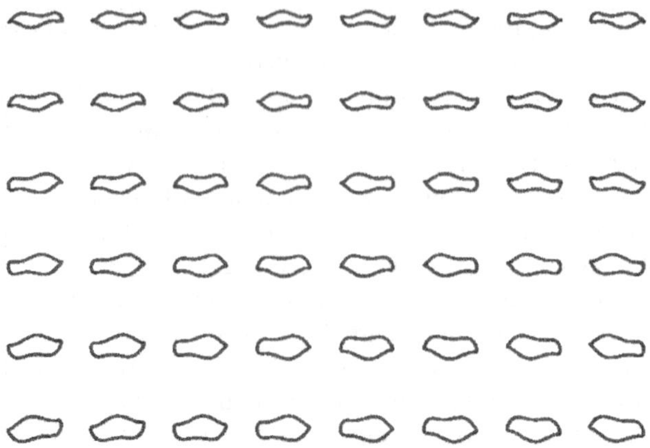

Fig. 12.16 A Kaluza–Klein wave disturbance moving through two of the ordinary dimensions

moving in our ordinary dimensions. So even though there is a standing wave in the extra-dimension, it is not standing still in our ordinary dimensions. The standing wave will be spread as a disturbance from one extra dimensional circle to another. In general we can imagine a propagating disturbance through an entire region of space as pictured in Fig. 12.16.

12.3 Einstein's Gravity

There is one other strange consequence of extra dimensions that we should know about, but we will first have to discuss Einstein's theory of gravity, which is also known as general relativity. You have probably experienced that during an elevator ride you feel heavier as the elevator first starts moving upward and lighter as the elevator first starts moving downward. In fact if the cables were cut and the elevator was falling freely you would feel weightless. If you dropped a ball it would not fall to the floor of the elevator, since you and the ball would both be falling toward the Earth, the ball would seem to float alongside you. Similarly, if you could ride in an elevator that was constantly accelerating upward in empty space you would feel like you were standing on the surface of a planet, feeling the tug of its gravitational pull. If someone sent a bullet or a pulse of light through a hole in the side of this elevator they would hit the far side of the elevator at a lower spot on the wall since the elevator would have accelerated up while the projectile crossed the elevator. The bullet or the light would be behaving like a projectile shot horizontally in the gravitational field of a planet: its trajectory curves down under the influence of gravity. The faster your elevator accelerates upward, the heavier you would feel, and the more the path of the bullet or the light would appear to be deflected. Thought experiments like these led Einstein to propose that light rays should be bent by gravitational fields and that we should think of gravity as a curving of space and time. The bending of light was later confirmed by comparing the positions of stars whose light skimmed past the Sun during a total solar eclipse with the positions of the same stars at any other time. In fact all the predictions of general relativity have been confirmed so far.

In Einstein's conception of gravity, mass and energy act as sources of curvature that bend space and time. The larger the mass or the energy the greater the curvature. Any object moving through space tends to follow the shortest path, which is not a straight line if the space is bent. This idea is familiar to anyone who has flown part way around the Earth. The shortest route between two points, the geodesic, is actually a section of a "great circle" which is obvious for two points on the equator but just as true for any other points. In general relativity a planet orbiting the Sun is analogous to a ball rolling on a bent rubber sheet, as pictured in Fig. 12.17.

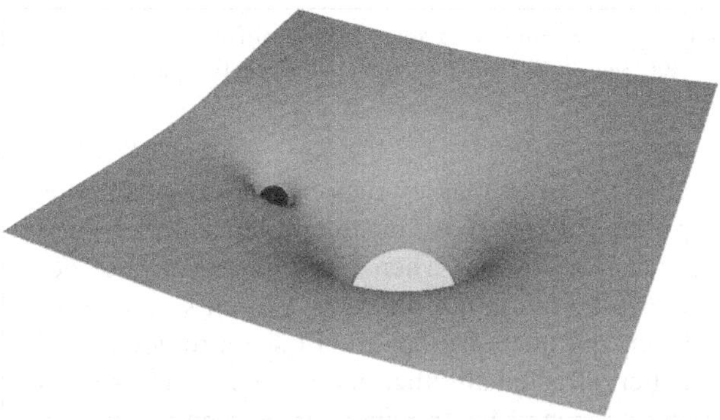

Fig. 12.17 A planet orbiting the Sun is analogous to a small ball on a rubber sheet rolling around the deformation caused by a large ball

Curved space may sound a little exotic but we can easily understand the basic point by thinking about some familiar objects. Euclid showed, a long time ago, that the sum of the angles inside a triangle always adds up to 180°, which is twice as much as the common 90° right angle. (For contrast, since a square is made from four right angles, the sum of the interior angles of a square is always 360°.) Of course, when Euclid said always, what he really meant was always if the triangle is on a flat piece of papyrus, or at least if the triangle is in a flat space. In a curved or non-Euclidean space, things are a little different. Imagine laying out a very large triangle on the surface of the Earth. Start with a 90° angle at the North Pole and project two lines (which are lines of longitude) down to the equator, and finish off the triangle with a segment of the equator (a line of latitude). Now the two lines of longitude make a 90° angle with the equator, so if we add up the angles inside our triangle we have 270°. If you tried the same kind of experiment on the surface of a saddle, you would find that the angles always add up to something less than 180°. Another way of seeing that curved spaces are different is that if you slice a globe representing the Earth from the North Pole to the South Pole and try to lay it flat to make a map without stretching it, then you will have to make more cuts as you may have seen in some types of maps that resemble flattened orange peels. If you try to take a saddle shape and flatten it you have the opposite problem that there is extra material that needs to be folded over (see Fig. 12.18).

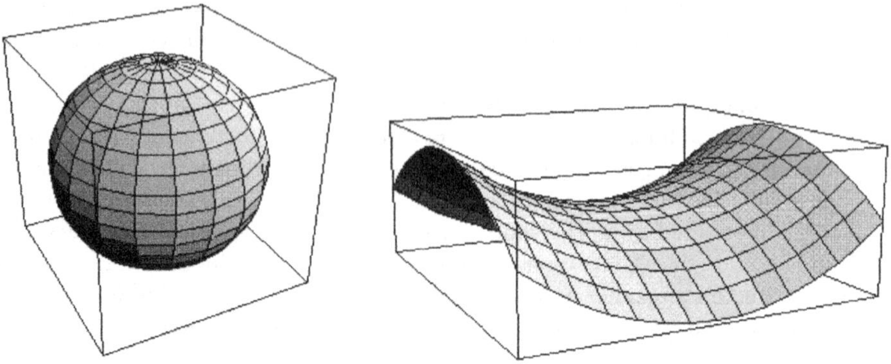

Fig. 12.18 A sphere is an example of a space with positive curvature since the angles inside a triangle add up to more than 180°, while a saddle has a negative curvature

In general relativity large masses distort space so that it is curved or non-Euclidean, and particles do not move in straight lines in such curved spaces.

An amusing side story is that when Einstein tried to use general relativity to describe our universe he ran into a problem. He felt sure that the universe should have existed forever, but his equations didn't seem to admit this possibility since all energy and matter attracted all other energy and matter, so things didn't want to stay put. A simpler version of his problem can be seen by throwing a ball up in the air. If you throw it hard enough it will go off into space, if you throw it normally it will go up a little way up and then fall to the ground. How do you throw it so it stops and hovers in mid-air? The short answer is that you can't. Einstein didn't let this stop him though, he changed his equations by adding an extra term, which he called the "cosmological constant." If this constant was positive it would effectively add a repulsion which Einstein thought could possibly balance the attractive force of gravity and give an unchanging universe. In 1929 Edwin Hubble showed that distant galaxies are rushing away from our own Milky Way galaxy and that the further away they are the faster they are moving. Faced with this, Einstein threw out his cosmological constant, calling it the "biggest blunder" of his career. Ironically recent astronomical studies suggest that our universe has something like a positive cosmological constant that is becoming more and more important as the universe grows older. A space dominated by a positive cosmological constant was first studied by Willem de Sitter in 1917, who showed that it makes the expansion of the universe actually speed up as opposed to slowing down as it would without a cosmological constant. So our universe seems to becoming more and more like a de Sitter space, although we now live in a more media savvy age, so the cause of the speed up is usually attributed to a mysterious "dark energy" rather than a boring old "cosmological constant."

Returning to extra dimensions we can now see that in a Kaluza–Klein theory there is a definite modification of gravity. In order to be consistent with general relativity the extra dimension must have the same kind of flexibility as our

4-dimensional space–time. As a consequence of curving the space, the size must be able to change. So we can imagine setting up traveling waves that are wiggles in the size of the extra dimension. If we imagine drawing just one of our ordinary dimensions along with the circular extra dimension we would have a tube. The waves we are talking about would be a moving wiggle in the radius of the tube as pictured in Fig. 12.19. Experimentalists have looked for the effects of these waves, but there has been no sign of them.

Fig. 12.19 A Kaluza–Klein graviton disturbance moving along one of the ordinary dimensions. This wave changes the size of the extra dimension

12.4 The Theory Formerly Known as String

The next significant advance in understanding possible extra dimensions had to wait for the development of string theory. String theory began as an attempt in the 1960s to understand the scattering of quarks that were tightly bound together in groups of three (as the proton[4] is) or pairs consisting of a quark and an anti-quark (these short-lived combinations of particles are called mesons). We now know that *gluons*, which are behind the strong force binding quarks, behave very differently from their cousins the photons that bind an electron to a proton. While photons do not interact directly with other photons, gluons do interact with gluons. The result is that the gluon field between two quarks tends to clump up into concentrated tubes rather than spreading in all directions as photons do. Viewed from a wide enough perspective these tubes can look like strings, so it is not too surprising in hindsight that people were led to thinking about strings when trying to understand the complicated behavior of quarks. In the 1970s the underlying role of gluons in the strong force became clear and a theory very similar to the one used for photons and electrons[5] was developed for gluons and quarks. As a result string theory was no longer considered very useful for understanding the scattering of quarks. In the meantime, people had found another reason for studying string theory: string theory seemed to automatically include an interesting particle called the graviton. A graviton is a

[4]The proton is the nucleus of the simplest atom, hydrogen, which also contains one electron hovering around the proton.

[5]The theory of photons and electrons is known by the formidable title of quantum electrodynamics or QED for short. The theory of gluons and quarks was, somewhat playfully, dubbed quantum chromodynamics or QCD.

particle that seems to be required for a quantum theory gravity, just as a photon is required for a quantum theory of electrodynamics. The basic idea of this new form of string theory was that all the known particles were really short pieces of some type of "string" at the most basic level, and that by understanding the theory properly one would be led to a unique theory of everything, including quantum gravity.

By the mid-1990s string theory had become very mathematically sophisticated. It was realized that for its own internal consistency string theory needed extra dimensions. Different versions of the theory had 10 or 11 dimensions in total which means 6 or 7 extra spatial dimensions. The fact that there was more than one "unique theory of everything" was somewhat embarrassing. In fact string theorists were working on five different string theories with esoteric sounding names: type I, type IIA, type IIB, heterotic $SO(32)$, heterotic $E_8 \times E_8$, and 11-dimensional supergravity. In fact the situation was somewhat akin to the fable of the blind men and the elephant. On encountering an elephant, each of the blind men touches a different part of the animal and pronounce their verdicts. Grabbing the tusk one says it is very much like a spear. The next feels the trunk and says it is very much like a snake. Another feels the leg and says it is very much like a tree, and so on (see Fig. 12.20). They are all

Fig. 12.20 The fable of *The Blind Men and the Elephant*. Illustration by Augusta Stevenson

partly right and at the same time all wrong, since they are failing to grasp the big picture.

Fortunately in 1995 Ed Witten was able to get a glimpse of the whole beast and he explained that each of the string theories people had been studying were actually a corner of a larger theory with 11 dimensions that he referred to as M-theory (see Fig. 12.21).

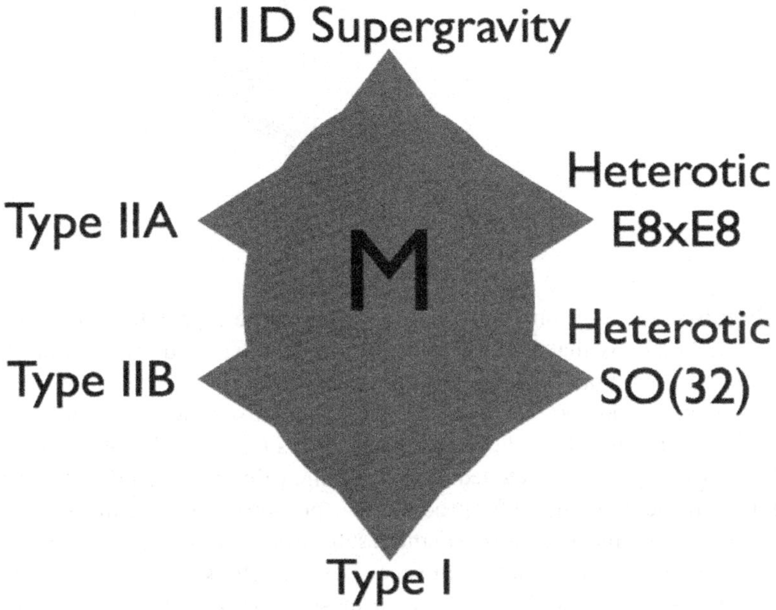

Fig. 12.21 Various string theories are thought to be special cases of overarching theory referred to as "M-theory"

Unfortunately, Witten did not explain why he chose the M in the name. Many suggestions have been put forward such as that it stands for "mother," or "magical," or "mystery," but my favorite is "membrane." That's because the new theory has a variety of membrane-like objects in it that are generalizations of ordinary 2-dimensional membranes (like the membrane that forms the head of a drum). These different types of membranes, or just "branes" for short, are distinguished by the number of spatial dimensions they fill. A 0-dimensional brane (or 0-brane) is just a point, which is what we might have been tempted to call a particle before M-theory. A 1-dimensional brane (or 1-brane) is a line, just like a string. A 2-dimensional brane (aka 2-brane) is a plane, while a 3-dimensional brane (3-brane) would fill up our entire 3-dimensional Universe, and a four dimensional brane (4-brane) fills one more spatial dimension than our Universe, and so on. It is very hard to draw a 3-brane or a higher dimensional brane on a 2-dimensional surface. Even a 2-brane poses something of a challenge since it is infinite, while any piece of paper is finite. Whenever I need to represent a 2 (or higher)-dimensional brane I will just draw a rectangle, like piece of paper, and ask you to imagine that it is really infinite (and

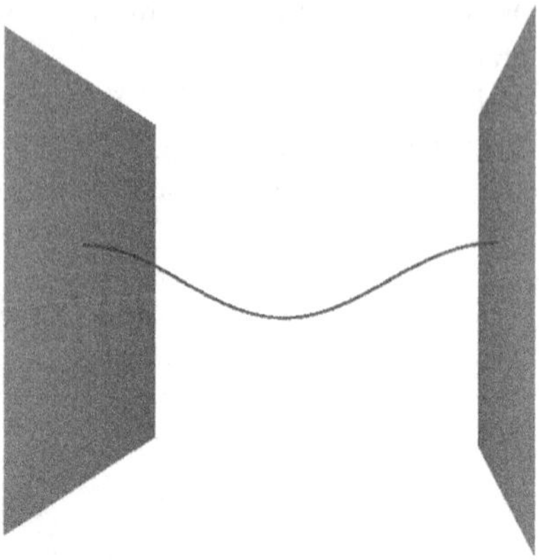

Fig. 12.22 Lower dimensional branes can end on higher dimensional branes. Here a string, or 1-brane, ends on two infinite higher dimensional branes where only two of the brane's spatial dimensions are shown

possibly fills some additional unseen dimensions). The especially interesting thing about these branes is that lower dimensional branes can end on a certain type of higher dimensional brane (known as D-branes); see Fig. 12.22.

When a string ends on two 3-branes, there is an associated minimum energy to produce such a string. This is because strings have a fixed amount of energy per unit length, so if the string is stretched along a straight line between the 3-branes then the minimum energy required for production is proportional to the distance between the branes. As with the Kaluza–Klein modes discussed earlier, this minimum energy is not associated with any motion in space, so when looked at with low resolving power instruments it would behave like a massive particle. Now as we move the 3-branes closer together, the string gets shorter, it takes less energy to produce, and so its mass gets smaller. If we take the distance between the branes to (zero), then we would expect the string would behave like a massless particle, like photon or a gluon. The string theory analysis finds that this is just what happens. There are even the correct number of independent photon polarization modes: 2. You can test this for photons using two polarized sunglasses. The polarizers in the glasses only let light pass that is polarized in one particular direction perpendicular to the direction that the photons are traveling. Putting another pair of polarizing sunglasses behind the first will still let light through, but rotating one pair by 90° will block all the light, since after the first pair the light is completely polarized and the second pair only lets through light polarized in a perpendicular direction. The massless string also has states that are like massless, chargeless electrons, which is a consequence of supersymmetry, but that is a story for another time.

With two 3-branes, strings can stretch from one brane to the other as I just described above, or it can go from one 3-brane back to the same 3-brane. Since there are two ways it can do this latter trick, there are three types of strings. With three 3-branes there are nine possible arrangements of a string. In general the number of possibilities grows like the square of the number of 3-branes (see Fig. 12.23). With more than one way to get massless strings one finds that the strings behave

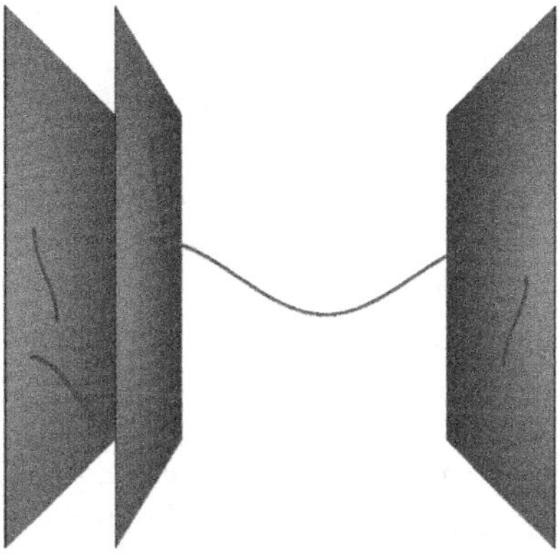

Fig. 12.23 With 2 branes a string can go from a brane to itself or from 1 brane to the other or start from the second and go to the first, so there are three types of strings. With 3 branes there are nine possibilities. In general with N branes there are $N \times N$ possible types of strings

like gluons rather than photons since they can interact with themselves directly (like gluons) while photons only interact directly with charged particles.

You may have been wondering why I have been insisting on using 3-branes in this last example rather than some other number. The reason is that when we have a stack of 3-branes all on top of each other they fill up a 3-dimensional universe, which could, after all, be our universe, or something very much like it. With the interacting gluon-like massless string states this theory does bear a resemblance to the type of particles and interactions we have seen in the real world. We can think of this theory, consisting of a stack of some number of 3-branes, as a toy model of the particles and interactions we find in the real world.

In the late 1990s, Juan Maldacena was studying this toy model universe and found that when he took the number of 3-branes to be very large things got much simpler rather than more complicated. He realized that he could think about this theory the way Einstein might have. The 3-branes are inside 6 extra spatial dimensions, and, since they have energy, they will bend the space around them according to Einstein's theory of general relativity. With a very large number of 3-branes, Einstein's equations are fairly easy to solve. When the dust settles the equations reveal 5 infinite dimensions. One of these dimensions is time, and three are the spatial dimensions of the 3-branes we started with (the dimensions where we could imagine living) which leaves 1 infinite extra dimension. The remaining 5 dimensions of the 6 extra dimensions that the 3-branes lived in turned out to have curled up into a 5-dimensional generalization of a sphere (a 5-sphere as opposed to an ordinary 2-sphere like the ones you see everyday) just as in a traditional Kaluza–Klein theory. With very low resolving power we wouldn't be able to see the curled-up dimensions, so this looks like a 5-dimensional gravity theory and it is relatively simple to understand.

What might seem confusing is that we now have two ways of looking at the same toy universe. The first way of looking at it is as a theory with gluon-like interactions. The gluon-like interactions are similar to those that bind quarks into protons; they are quite difficult to understand since the most practical way to deal with them is a giant super-computer simulation. At least they only involve three spatial dimensions though. The second way of looking at this toy universe is Maldacena's gravity description with 4 spatial dimensions where one can figure out how things work with a few equations, pencils, and paper (I might ask for an eraser too). Now, these two ways of looking at things might seem, at first sight, to be completely contradictory. They don't even agree on how many dimensions there are. Maldacena showed, however (and many others have extended his results since), that in every way that can be checked they do agree. This led to the celebrated Maldacena conjecture that says, roughly speaking, that a particular 4-dimensional theory with strong gluon-like interactions is really equivalent to gravity in a particular 5-dimensional space. I say "a particular 5-dimensional space" because it is not a flat space, but a space with a negative curvature and a negative cosmological constant. Since the space with a positive cosmological constant is called de Sitter space, this special type of curved (or warped) space is called anti-de Sitter space. The reason that the two different theories can agree is that the 4-dimensional theory turns out to look the same when viewed at any length scale. More precisely if I doubled the size of my length and time units (inches and seconds say) and halved the size of my units of mass and energy, then every prediction of the theory is unchanged. The same thing happens in the 5-dimensional theory with the twist that doubling the length unit of the extra dimension is the same as halving the energy.

12.5 Warped Extra Dimensions

Typically in the past when someone tried to make a string theory that resembles the real world, the extra 6-spatial dimensions got curled up in something called a Calabi–Yau manifold which is much more complex than a 6-dimensional sphere. Sometimes such a Calabi–Yau manifold will have one or more long thin protuberances (or "throats") sticking out of it, as in Fig. 12.25. Inside of these throats the space can look very much like an anti-de Sitter space along with some additional small curled-up dimensions. If the throat is much longer than it is wide it can be approximated by a line, or in other words the throat behaves like there is just 1 extra dimension. This is because the largest extra dimension would be detected in an experiment long before the other extra dimensions since they are so much smaller and thus much harder to resolve. The simplest case is a single throat that looks like an anti-de Sitter space with a finite length. In 1999 Lisa Randall and Raman Sundrum pointed out that if there was an extra dimension like this it could solve one of the deepest mysteries in our present understanding of the universe: the origin of mass.

By "origin of mass" I mean how is it that an electron has a mass while a photon is massless. In fact there are other force carrier particles called the W and Z that are very similar to the photon except for the fact that they have very large masses. The W and Z are responsible for the so-called weak force, which is weak compared to electromagnetism precisely because the force carrier particles have a mass. A

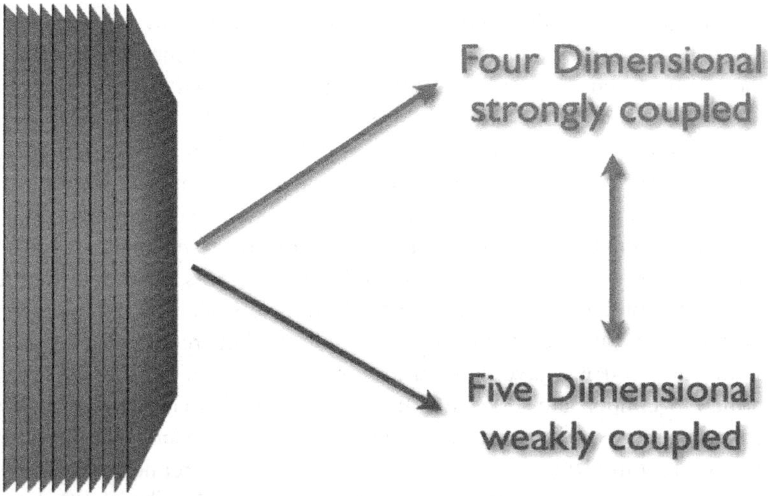

Fig. 12.24 Maldacena found two descriptions of a stack of branes, one with strongly coupled gluons in 4-dimensions and one with weakly coupled gravity in 5-dimensions. Since the two seemingly contradictory descriptions describe the same thing, they must be equivalent

Fig. 12.25 Multiple Randall–Sundrum "throats" hanging out from a compact Calabi–Yau manifold. If we probe this space with waves with peak to peak distances around the size of the length of the throat, then each throat looks like a single warped extra dimension

further mystery is that all of the experimental data we have suggest that in the early universe, just after the Big Bang, the *W* and *Z* were massless just like the photon. In fact even the electrons and quarks were massless, but somehow, as the universe cooled down, electrons, quarks, *W*'s, and *Z*'s all got masses.

To get an idea of how this might be possible it is helpful to know a little about superconductors. Superconductors are very useful for technological applications since they allow a current to flow without resistance. Usually materials have to be cooled to very low temperatures in order for them to become superconducting. Figure 12.26 shows what happens around a cube of metal as the temperature is lowered from a high temperature (where it behaves like a normal metal) to a low temperature where it is superconducting. The metal cube is surrounded by vertical lines which represent a uniform magnetic field. When the temperature is low enough for the metal to be superconducting it is impossible for this magnetic field to exist inside the superconductor. In other words the magnetic field is expelled from the superconductor. What's happening is that inside the superconductor a large number of pairs of electrons have entered a coherent quantum state (they are said to have formed a "condensate"). What is important for us is that the condensate gives photons inside the superconductor an effective mass. This means that there is a minimum energy required for a photon to exist inside the superconductor. This is why the magnetic field can't exist inside the superconductor, the photons that make up the magnetic field have too little energy. Einstein's formula requires a minimum energy given by the effective mass times the speed of light squared.

We believe that something similar happens in our universe for the weak interactions. As the universe cooled a condensate formed that forced the W and Z particles to have non-zero masses. The difference between this process and ordinary superconductors is that this "weak condensate" couples to W and Z particles rather than photons and it permeates all of space. The simplest model that describes how this can happen is called the standard model of particle physics. In the standard model there is a Higgs field (named after Peter Higgs) that forms a weak condensate and gives the W and Z their masses. In fact the standard model supposes that electrons and quarks get different masses depending on how much they interact with the Higgs field. Those particles that interact very little with the Higgs field (like the electron) get small masses, while those that interact more strongly (like the top quark) get a very large mass. It is somewhat like people wading through a pool of molasses. A person with long legs who is only in up to their knees can move much more easily than a shorter person who is in up to their waist.

The telltale sign that the standard model story is correct is that we should see the ripples in the Higgs field: that is a particle known as the Higgs. There have been many experimental attempts to find the Higgs particle, but so far no one has seen

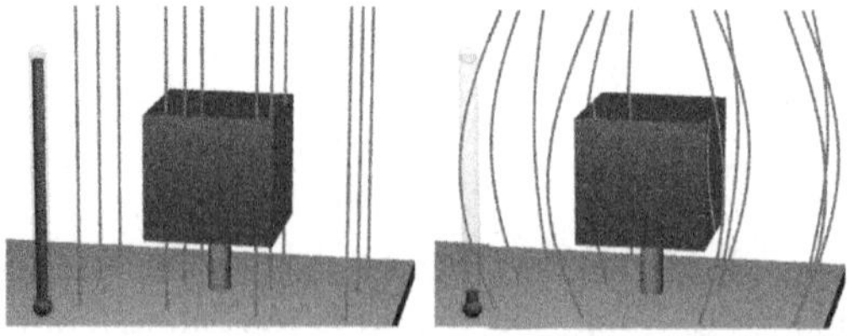

Fig. 12.26 Imagine that we could see magnetic field lines around a superconducting metal. At high temperatures the magnetic field penetrates the metal but below a certain temperature the magnetic field is pushed outside of the metal

direct evidence for it. So, around the turn of the 21st century physicists believed that space was filled with a "Higgs condensate," but no one could detect any sign of it. The upcoming Large Hadron Collider should turn on this year (2009) and will have enough energy to produce any variant of the Higgs particle consistent with the standard model of particle physics: the Large Hadron Collider will provide a definitive test of the standard model. Will we see a Higgs? Or will the Higgs condensate go the way of the aether? Only time will tell.

Even if we do find a Higgs particle at the Large Hadron Collider, most physicists expect that there will also be other new particles that are light enough to be seen at the Large Hadron Collider. Even though the standard model has already been tested in various ways at the fraction of a percent level (and passed with flying colors) it does suffer from a serious problem. When people calculate the quantum corrections (that is the corrections that take into account quantum mechanics) to the standard model they find that the corrections tend to drive the Higgs condensate up to the highest scale in the theory, which gives *W* and *Z* masses about a million trillion times bigger than the value found in experiments. Technically this problem can be overcome by adjusting one of the basic numbers in the theory to an accuracy of 32 digits. This is called "fine-tuning" and usually causes a deep feeling of revulsion in a physicist. It would be like owning a radio whose tuning knob had to be adjusted very precisely in order to get your favorite radio station. If the knob on your radio had to be adjusted to a precise angle down to a million-trillionth of a degree you would probably just throw the radio away since you would be extremely unlikely to ever be able to tune your radio precisely enough to listen to your favorite station. Using the same kind of logic most physicists have thrown away the standard model in the sense that they expect that the Large Hadron Collider will prove it wrong, even though it is perfectly adequate to describe any other particle physics experiment that has ever been done.

The situation in particle physics is actually slightly stranger than with the impossibly-difficult-to-tune radio since we have already seen the *W* and *Z* and know what their masses are. It is more like we are already listening to the radio station and wondering how the radio got tuned so accurately. Perhaps an even better analogy is that you walk into a room and find a pencil standing on its tip. Now it is possible that someone so carefully balanced the pencil that it is still standing and will eventually fall, but if you actually saw this I suspect that you would tend to doubt that explanation. You might start poking around the room trying find out if there are some very thin wires holding the pencil up, or perhaps a magnetic field, or even a little dab of super-glue. This is what theoretical physicists have been doing for many years now: trying to come up with some explanation of how the *W* and *Z* masses are kept stable at their observed values. These explanations almost always end up having new particles that we should be able to see at the Large Hadron Collider, so most particle physicists not only think that we will find the Higgs particle but also find out what stabilizes the *W* and *Z* mass scale.

One possibility for stabilizing the masses of the *W* and *Z* particles is the extra-dimensional model proposed by Lisa Randall and Raman Sundrum. In their model the extra dimension is warped, just as it is in the anti-de Sitter space of Maldacena's toy universe sandwiched between two 3-branes. Because of this warping, the length of an object that moves through the extra dimension changes, getting larger as it moves in one direction and smaller in the other (see Fig. 12.27). This also means that the length between wiggles of a wave changes in the same way. As the length between wiggles gets larger, the frequency gets smaller, and therefore since energy is proportional to frequency, the energy gets smaller. In the Randall–Sundrum model the Higgs condensate only exists on the brane at the end of the extra dimension where the energies are the smallest. If the highest energy on this brane is not too much larger than the rest energy of a *Z*, then there will be no problem stabilizing the *Z* mass.

We can take this model a step further and consider what happens if we allow the highest energy on the brane to be much larger. What we expect is that this will make the quantum corrections to the Higgs condensate and thus the *Z* mass on this brane larger. But if the *Z* is free to move through the extra dimension then it will tend to avoid this brane since it costs a lot of energy for it to be on the brane. As we raise the highest energy on the brane, it is not too long before the *Z* particles are practically never on the brane. If we think about what this means for the standing Kaluza–Klein waves we will find that standing waves have to be 0 on the brane where the *Z* particles never appear. So we will find standing Kaluza–Klein waves that look like the right-hand column of waves in Fig. 12.28. The lowest energy wave possible under these circumstances is shown in the top right figure with just one quarter of a full wiggle. However, since the photon does not interact with the Higgs condensate it can have a constant standing wave (as in the top-left wave in Fig. 12.28) since it isn't forced to be zero anywhere. The photon can have a wave with no wiggles along the extra dimension, so the photon remains a massless particle, since there is no minimum energy required to produce it. In this type of universe, since the Higgs condensate is large, the particles that interact with the Higgs get large masses on the brane where the Higgs is, so like the *W* and *Z* particles they avoid this brane. In fact the ordinary particles we know and love would not be able to produce a Higgs particle in a collision, since they either don't interact with the Higgs or are never in the place where Higgs particles live. By raising the Higgs condensate while stabilizing the *Z* mass scale we have effectively removed the possibility of experimentally studying the Higgs, so this type of model is called "Higgsless".

Returning to the Randall–Sundrum model, the fact that the highest energy on the brane is not much above the *W* rest energy means that this model will soon be tested at the Large Hadron Collider. Even the Higgsless type of model, and many others as well, will be tested at this new machine.

12.6 How Do We Look for Extra Dimensions?

It might seem that the prospects for experimentally verifying the Higgsless model are pretty dim. Since any possible extra dimensions are probably much smaller than an atom it could seem practically impossible to find any evidence of them. Actually,

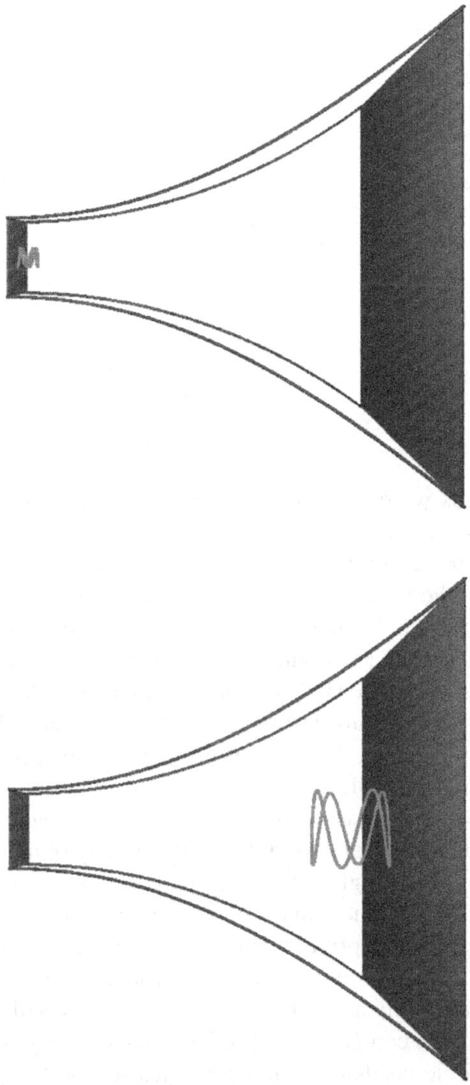

Fig. 12.27 In the Randall–Sundrum model, the extra dimension is warped, so lengths change as one moves through the extra dimension. A wave that moves through the extra dimension has a different distance between peaks when it completes its journey

evidence for extra-dimensional models, like the Higgsless scenario, which potentially solve the fine-tuning problem of the standard model is usually not too hard to find. Since they need to play a role in stabilizing the *Z* particle mass scale there have to be new particles that we haven't seen before which are not much heavier than the *Z*. Typically they are no more than about 10 times as heavy as the *Z*. For models

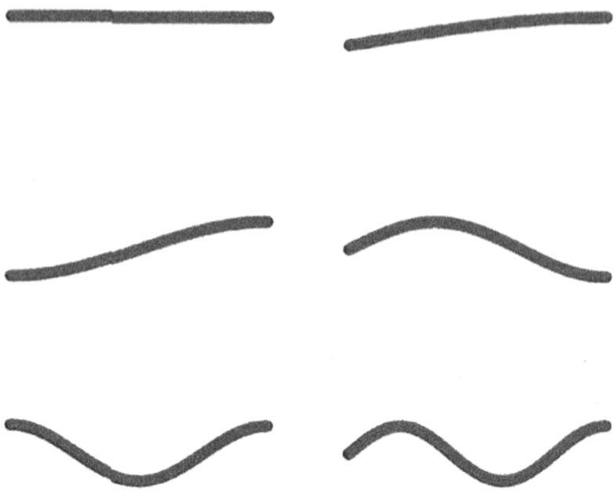

Fig. 12.28 The figures on the *left* show standing waves where both ends are free to wiggle. The *top left* figure is a constant standing wave. The figures on the *right* show standing waves with the *left end* forced not to wiggle and the other end free to wiggle

with extra dimensions we can use the relation between the masses of the Kaluza–Klein standing waves and the length of their wiggles in the extra dimension to find the size of the required extra dimension. This is because the lightest Kaluza–Klein standing wave (with the fewest wiggles) has a wiggle about the length of the extra dimension. So for masses 10 times heavier than the *Z* particle, we find a size for the extra dimension around a thousandth of the width of a proton. This is just in the range that the upcoming Large Hadron Collider can probe. As I have said the Large Hadron Collider is scheduled to start running later this year (2009)!

It is not just a coincidence that this experiment will be probing the right length scales. The Large Hadron Collider was designed to find out exactly how particles (including the *Z*) got their masses, so it will have a top energy of 150 times the energy of a *Z* particle at rest, or 14,000 times the energy of a proton at rest. Hadron is the name given to all strongly interacting particles like protons and mesons that are made out of quarks. As the name implies, the collider will have two beams of hadrons (which are actually protons since they are the easiest type of hadron to work with) being steered by magnets around a roughly circular tunnel in opposite directions. At certain points around the circle the beams will be steered into each other so that the protons can collide. In order to have a beam of protons with such large energies the circle needs to be about 5 miles across. Each proton in the beam will have 7000 times as much energy of motion as it has rest energy, so if two protons meet in a perfect head-on collision there is enough energy to make a single particle whose mass is 150 times the mass of the *Z*. In a more typical collision there will be enough energy to produce a particle with 50 times the mass of the *Z*, so we should be able to see the new particles that help the *Z* get its mass and whether extra dimensions play any role.

Of course it's not quite as simple at just staring really hard at the collision point, trying not to miss anything. When the protons collide we expect to see big sprays of many particles shooting out in many directions at once. Enormous particle detectors, called ATLAS and CMS, have been built to capture as much information about as many of these particles as possible. ATLAS is about 75 ft (25 m) high, about the

Fig. 12.29 The ATLAS detector at the Large Hadron Collider. The tiny things at the *bottom* of the picture, just *left of center*, are people

Fig. 12.30 The CMS detector at the Large Hadron Collider

height of a seven-story building. CMS, is smaller (the C in CMS stands for compact) and is only about 45 ft high. These detectors were designed to contain the mini-explosions created by colliding protons and to record the details of which particles were produced, how much energy they had, and which direction they were moving. In a sense, the particle detectors will be taking high-speed digital snapshots of the blast that results when the energy of the protons is converted into producing new particles.

An example of the kinds of things that can happen at the Large Hadron Collider is shown in Fig. 12.31. Let's imagine what things would look like if particles left trails behind them as they moved through space. Parts of the ATLAS and CMS detectors called *tracking chambers* (or trackers) can record such trails for electrically charged particles, but they cannot do this for electrically neutral particles. Also real trackers do not have enough resolution to pick out individual quarks or particles that only live for a very short time. Supposing we had this perfect tracking ability, let us just consider one quark in a particular proton in one of the beams and a second quark in a proton in the other beam. After turning on our tracking ability for a short amount of time we might see something like part (a) of Fig. 12.31. In the short amount of time the tracking has been on, the quarks have moved a little and left two short tracks. A little later in (b) the quarks have moved further and in the meantime both have emitted a Z particle. Since the quarks have given some of their momentum to the Z's, their directions have changed slightly. Later still in (c) the Z particles have collided and annihilated, so that now there are two quarks and a Higgs particle. In (d) the Higgs particle moved a little before reversing the previous process and turning back into two new Z's. One usually refers to a process where a single particle turns into two or more particles as a decay, so we could say the Higgs has decayed into two Z's. Later on in (e) one of the Z's has decayed into an electron and a positron. Since the Z has no electric charge, it must decay into particles whose charges balance out to zero charge when they are added up. Finally in (f) the last Z has also decayed to an electron–positron pair.

Experimentalists at the Large Hadron Collider will, of course, not have the luxury of seeing such fine detail. They will be able to see the tracks left behind by particles like the electrons and positrons; by measuring their energies one could start to reconstruct what the mass of the Higgs particle had to be. The detectors will also record the sprays of particles that would be produced by each of the quarks after their strong interactions with gluons are taken into account. We have neglected the other quarks and gluons in each of the protons that the two quarks we tracked were inside of. There would be a plethora of other particles that appear in the detector as a result of the strong interactions between all the quarks and gluons. The extra complication of all these additional particles from the quarks and gluons makes it more difficult for the experimentalists to understand what is going on in their experiment. This is the price of using a hadron collider. The reason that experimentalists are willing to put up with these difficulties is that hadron colliders are much cheaper to build than electron (and/or positron) colliders with the same energy and collision rate.

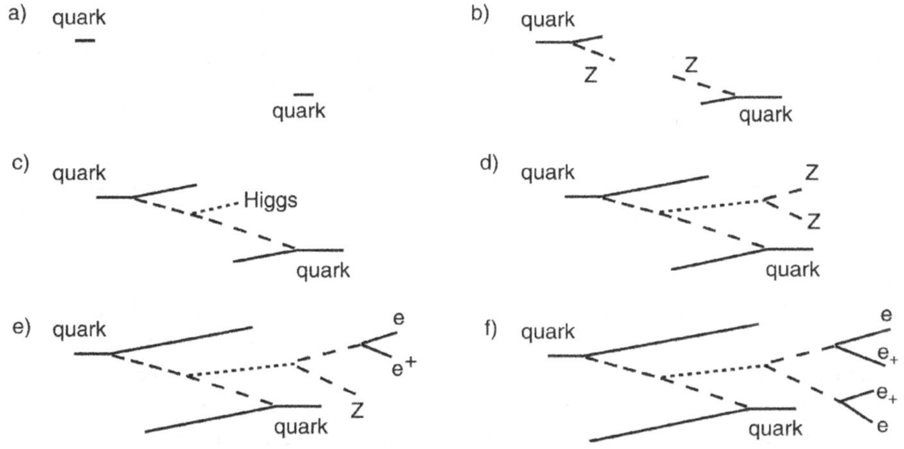

Fig. 12.31 Imagine that particles left trails behind them so that we could follow their progress. In (**a**) we see two quarks approaching each other. In (**b**) each quark has emitted a Z particle. In (**c**) the Z particles have collided and converted into a single Higgs particle. In (**d**) the Higgs particle has turned back into two new Z particles. In (**e**) one of the Z particles has decayed into an electron, labeled e, and a positron, labeled e^+. In (**f**) the second Z particle has decayed in the same fashion

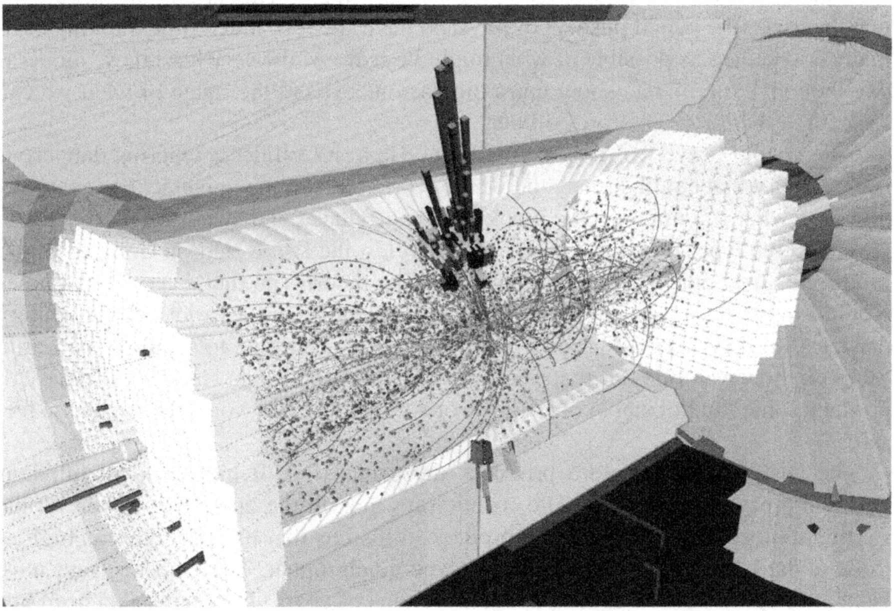

Fig. 12.32 Spays of particles in a computer simulation of a proton–proton collision event in the CMS detector. The computer model was generated by Matt Strassler with help from Peter Skands, and processed through a CMS detector simulation by Albert de Roeck, Christophe Saout, and Joanna Weng. Visualized by Ianna Osborne

In addition to trackers, the detectors will have *calorimeters* that measure the total energy that comes out in any particular direction. The name suggests a tool for counting calories, which are a unit of energy after all. These devices are very useful for looking for Kaluza–Klein particles that can disappear into an extra dimension. When a particle disappears into an extra dimension it takes some energy and momentum with it, which means that it will look like some energy and momentum was lost in the collision. When this happens we have a so-called missing energy event. Such events stick out like a sore thumb (or a missing thumb) so they are relatively easy to find. An historical example of how this strategy works is the case of the neutrino. Problematic decays of atomic nuclei, where it seemed energy was getting lost, lead Wolfgang Pauli to propose in 1930 an almost invisible particle that became known as the neutrino. It took until 1956 before Fred Reines and Clyde Cowan were able to experimentally verify the existence of neutrinos.

If the Large Hadron Collider works as designed, there will be about a billion collisions per second. Even using the latest computers and electronics, the detectors will not be able to keep up with that pace. They will only be able to record about 100 events per second for further analysis. Computers inside the detectors have to decide in a ten millionth of a second which collisions look interesting enough to save. For this reason it is important that the experimentalists who built the detector have a pretty good idea of what they were looking for, if they didn't, the computer might miss the really interesting events and not record them. This is one of the reasons that theorists (the pencil pushers of physics) have spent so much time thinking of as many crazy ideas as possible of what might be seen. At the very least these modern-day thought experiments about extra dimensions expand the range of what we can look for at the Large Hadron Collider.

Some people have worried that the Large Hadron Collider is entering dangerous new territory that could destroy the whole world or even the universe. For example, they say, what if a microscopic black hole is produced and swallows up the Earth. Now in some types of extra-dimensional theories, the Large Hadron Collider would produce microscopic black holes, but these same theories predict that the black holes would evaporate in a fraction of a second by spewing out a variety of elementary particles, so they would disappear long before they had time to swallow the Earth. Suppose, however, that our theories are wrong, and the Large Hadron Collider will produce some kind of stable black hole, or something even worse. How can we feel safe?

The answer is that nature produces the same kind of high-energy collisions between protons all the time. The cosmic rays that hit the upper atmosphere of the Earth are mostly protons and they produce collisions that have energies as high as those at the Large Hadron Collider and even much higher. These cosmic rays have been striking the Earth's atmosphere for billions of years. They have also been hitting the surface of the Moon, the stars, and even things like neutron stars. The fact that we, the Moon, and neutron stars have not been swallowed up over these billions of years tells us that the Large Hadron Collider should be perfectly safe.

On September 10, 2008, the Large Hadron Collider began its final testing phase. With much media fanfare, proton beams were successfully sent around the huge ring. Unfortunately, after only 9 days an electrical fault melted a hole in the liquid helium cooling system. The force of the rupture ripped the beam-line support pillars out of the concrete floor. This was not the kind of mini-explosion we were expecting! It will take until June 2009 before the repairs are completed. Hopefully by then all the bugs will be worked out and we will start to get some new data and some answers. For physicists who have been waiting almost 20 years for this experiment, a few more months won't make too much difference.

12.7 Epilogue

Extra dimensions have fascinated people for over 100 years, stimulating new ideas about the nature of reality and new models of how the universe works. The hot problem of particle physics research now is the origin of the particles' masses and the associated search for the Higgs particle. Our best current theory of how masses arise, the standard model, has been tested to high accuracy in many respects, but it has a problem. Quantum physics seems to make the predicted masses in this theory more unstable than a house of cards, even more unstable than a pencil standing on its tip. If our universe has small extra dimensions, around a thousandth of a proton wide, they could be the key to stabilizing our universe at these tiny scales. Various models of extra dimensions have given us new possible answers to this stability question, and in a few years the Large Hadron Collider should tell us nature's answer.

References

1. E. A. Abbott "Flatland: A Romance of Many Dimensions," Princeton University Press, Princeton (1884).
2. F. Jerome "The Einstein File: J. Edgar Hoover's Secret War Against the World's Most Famous Scientist," St. Martin's Press, New York (2002).
3. B. Mahon "The Man Who Changed Everything: The Life of James Clerk Maxwell," Wiley, New York (2004).
4. D. Hambling "Let the light shine in," The Guardian (UK May 30, 2002), http://education.guardian.co.uk/higher/medicalscience/story/0,,724257,00.html.
5. L. Randall "Warped Passages: Unraveling the Mysteries of the Universe's Hidden Dimensions," Ecco, New York (2005).
6. L. Krauss "Hiding in the Mirror: The Quest for Alternate Realities, from Plato to String Theory (by way of Alice in Wonderland, Einstein, and The Twilight Zone)," Penguin, New York (2006).

Chapter 13
Extra Material: The Equations Behind the Words

13.1 Units and Coordinates

It is usually more convenient to use metric units (technically the International System of Units or SI). A brief list of some of the more useful units is given in Table 13.1. A meter is about a yard long and a raisin weighs about a gram. Water boils at 373 K, room temperature is about 293 K, water freezes at 273 K, and the remnants of the Big Bang explosion are now at about 3 K. Nothing can be colder than absolute zero, which is 0 K.

One of the advantages of using SI is that when going from large to small (or vice versa) it is easier to change to a more convenient unit, since larger units are just multiples of 10 times the basic unit, see Table 13.2. For example 1 km is 1000 m and 1 m is 100 cm, so a kilometer is 100,000 cm and a kilogram is 100,000 cg. Now, quickly, a mile is how many inches?

Of course, some units can be expressed in terms of more basic units. We know that a velocity being distance traveled divided by the time taken could be measured in meters per second (m/s) and an acceleration being the rate of change of velocity could be measured in m/s^2. Furthermore Newton taught us that force equals mass times acceleration, so a force is measured in kg m/s^2, and energy can be thought of in terms of work, which is the amount of forces exerted through a distance, so in fact, 1 J is equal to 1 kg m^2/s^2. We could also have arrived at the units of energy from Einstein's relation energy equals mass times the speed of light squared. Keeping track of the dimensions of various quantities in an equation is a good way of avoiding mistakes. Sometimes just knowing the dimensions of some quantities allows you to estimate the value of another unknown quantity (without knowing the exact formula that relates them) just by arranging things so the dimensions match up correctly. This trick, which goes by the fancy name of *dimensional analysis*, was another of James Clerk Maxwell's innovations. While technically electronvolt is not an SI unit, it is widely used in particle physics since it is much more convenient to use for the small amount of energy a single particle typically has. One electronvolt is equal to 1.602×10^{-19} J.

There are several equivalent ways of specifying a point in a 4-dimensional space–time. We can simply give the values of the coordinates along three perpendicular

I. Bars, J. Terning, Founding Editor: F. Nekoogar, *Extra Dimensions in Space and Time*, Multiversal Journeys, DOI 10.1007/978-0-387-77638-5_13,

Table 13.1 SI units for different quantities

Quantity	Unit	Abbreviation
Length	meter	m
Time	second	s
Mass	gram	g
Temperature	kelvin	K
Frequency	hertz	Hz
Energy	joules	J
Energy	electron volt	eV

Table 13.2 Prefixes and corresponding powers of 10 for SI units

Power of 10	Prefix	Abbreviation	Number
10^{15}	peta	P	1,000,000,000,000,000
10^{12}	tera	T	1,000,000,000,000
10^{9}	giga	G	1,000,000,000
10^{6}	mega	M	1,000,000
10^{3}	kilo	k	1,000
10^{2}	hecto	h	100
10^{1}	deca	da	10
10^{-1}	deci	d	0.1
10^{-2}	centi	c	0.01
10^{-3}	milli	m	0.001
10^{-6}	micro	μ	0.000001
10^{-9}	nano	n	0.000000001
10^{-12}	pico	p	0.000000000001
10^{-15}	femto	f	0.000000000000001

directions (e.g., east–west, north–south, and up–down) as well as the time. Often these coordinates are grouped in a vector, so 5 km east, 4 km north, 1 km high, and 2 o'clock could be written as a *vector*

$$(2\,\text{h}, 5\,\text{km}, 4\,\text{km}, 1\,\text{km}), \tag{1}$$

where we have chosen (arbitrarily) to write the time first. We could also talk about a position in space–time by referring to the four numbers in the vector as four variables called x_μ where the Greek letter μ (sounds like "mew") runs over four different integer values to distinguish the different pieces. For historical reasons the spatial coordinates are labeled by $\mu = 1, 2, 3$ and the time component is often labeled by $\mu = 0$. So in the example above we would write

$$x_0 = 2\,\text{h}, \quad x_1 = 5\,\text{km}, \quad x_2 = 4\,\text{km}, \quad x_3 = 1\,\text{km}. \tag{2}$$

To be completely clear one could also call the spatial components x, y, and z and the time t, so our example would be written as

$$t = 2\,\text{h}, \quad x = 5\,\text{km}, \quad y = 4\,\text{km}, \quad z = 1\,\text{km}. \tag{3}$$

13.2 Einstein and the Fourth Dimension

In the example of rotating the map (see Fig. 13.8) we used the fact that the distance between two points doesn't change if we rotate either the Earth or our choice of coordinates. We say that lengths are *invariant* under rotations. In our specific example suppose the road running east–west running beside the house has kilometerage markers and the house is at the 38 km marker and the corner is at the 41 km marker. If we use the variable x to describe the value of the kilometerage marker, then in equations we can write

$$x_{\text{house}} = 38\,\text{km}, \quad x_{\text{corner}} = 41\,\text{km}. \tag{4}$$

We can use the Greek letter Δ (pronounced "delta"), in order to simplify the equations we will come to shortly. We use Δ to remind ourselves that Δx is the difference between two numbers, in this case the difference in two numbers specifying positions along a certain direction. The distance to the east was 3 km and the distance to the north was 4 km. In equations we can write the distance between the corner and the house as

$$\Delta x = x_{\text{corner}} - x_{\text{house}} = 3\,\text{km}. \tag{5}$$

Similarly along the north–south road the distance between the corner and the office is

$$\Delta y = y_{\text{office}} - x_{\text{corner}} = 26 - 22\,\text{km} = 4\,\text{km}. \tag{6}$$

Finally thanks to Pythagoras we know that the distance from the house to the office is 5 km, since

$$(5\,\text{km})^2 = (\Delta x)^2 + (\Delta y)^2 = 9\,\text{km}^2 + 16\,\text{km}^2, \tag{7}$$

where the superscript 2 (or squared) means "multiply something by itself." Since we have multiplied two lengths, the answer has units of an area, square kilometers.

After we rotate the directions that we call north and east, there are new values for the length of the sides of the triangle with one side in the new east direction, one side in the new north direction, and one side going from the house to the office. We can call the kilometerage along the new east direction x' where the superscript ($'$) tells us that this is using our new *coordinate system*. In our example we changed north by 41.5931° so we have[1]

$$\Delta x' = \cos(41.5931)\,\Delta x + \sin(41.5931)\,\Delta y = 4.89898\,\text{km}, \tag{8}$$

$$\Delta y' = -\sin(41.5931)\,\Delta x + \cos(41.5931)\,\Delta y = 1\,\text{km}. \tag{9}$$

---

[1] If you are not familiar with the trigonometric functions cosine and sine (cos and sin for short) that's all right, they are the mathematical tools that relate rotation angles to coordinate changes.

But the house and the office haven't moved, we only changed what we called north and east, so the distance between the house and the office can't have changed. Indeed we find

$$(5\,\mathrm{km})^2 = (\Delta x')^2 + (\Delta y')^2 = 24\,\mathrm{km}^2 + 1\,\mathrm{km}^2. \tag{10}$$

For a rotation by an angle, represented by the Greek letter θ (pronounced "theta"), we would find

$$(5\,\mathrm{km})^2 = (\Delta x')^2 + (\Delta y')^2 \tag{11}$$

$$= (\cos(\theta)\Delta x + \sin(\theta)\Delta y)^2 + (-\sin(\theta)\Delta x + \cos(\theta)\Delta y)^2 \tag{12}$$

$$= (\cos^2(\theta)\Delta x^2 + \sin^2(\theta)\Delta y^2 + 2\cos(\theta)\sin(\theta)\Delta x\Delta y) \tag{13}$$

$$+ (\sin^2(\theta)\Delta x^2 + \cos^2(\theta)\Delta y^2 - 2\cos(\theta)\sin(\theta)\Delta x\Delta y) \tag{14}$$

$$= (\cos^2(\theta) + \sin^2(\theta))(\Delta x^2 + \Delta y^2) \tag{15}$$

$$= \Delta x^2 + \Delta y^2, \tag{16}$$

where we have used the fact from trigonometry that $\cos^2(\theta) + \sin^2(\theta) = 1$ for any angle θ.

We can easily extend this discussion to 3 dimensions of space. In everyday language we refer to the three spatial directions as forward/backward, left/right, and up/down, but mathematicians and physicists often call these directions x, y, and z, and specify spatial positions by the distances along each of these directions measured from some convenient reference point, which is often called the *origin*. So a physicist would locate his house at $x_{\mathrm{house}}, y_{\mathrm{house}}, z_{\mathrm{house}}$, which gives new meaning to the phrase "location, location, location." For rotations in 3 dimensions the invariant length L is

$$(\Delta L)^2 = (\Delta x)^2 + (\Delta y)^2 + (\Delta z)^2, \tag{17}$$

where Δx, Δy, and Δz are the distances along the three mutually perpendicular directions measured from the point we are rotating around.

Einstein explained to us if we want to consider a fourth dimension, time, that there is another type of coordinate change that we need to be careful about, that is changing velocities. Changing the velocity of our coordinate system is called a *boost*, and just as a rotation leaves the length ΔL in Eq. (17) invariant, a boost leaves the following quantity invariant:

$$(\Delta s)^2 = c^2(\Delta t)^2 - (\Delta x)^2 - (\Delta y)^2 - (\Delta z)^2, \tag{18}$$

where Δt is a time interval and c is the speed of light. We could have multiplied this equation by -1, so that the terms with spatial differences contributed positively, and still found an invariant. The overall sign doesn't matter, as long as we stick to our choice consistently, what matters is the relative minus sign between the time contribution and the spatial contribution. The speed of light is measured to be

$$c = 3.0 \times 10^8 \text{ m/s}, \tag{19}$$

or 186,000 miles/s. Note that the c^2 in the first term in Eq. (18) when multiplying $(\Delta t)^2$ makes this term have the same units as the other terms: length squared. Now consider two *events*, call them 1 and 2, that are separated by a distance

$$\Delta x = x_2 - x_1 \tag{20}$$

in the Easterly (or x) direction[2] and by a time interval

$$\Delta t = t_2 - t_1. \tag{21}$$

In other words event 1 is at position x_1 when a clock at that position reads t_1 and event 2 is at x_2 when the clock at that position reads t_2. In this example Δy and Δz are zero since both events take place at the same distance along the y-direction and same height in the z-direction. So the relativistic invariant separation in this example is

$$(\Delta s)^2 = c^2(\Delta t)^2 - (\Delta x)^2. \tag{22}$$

If we start moving in the x-direction with a velocity v (that is we "boost" to new coordinates that are moving with a velocity v in the x-direction relative to our old coordinates), then we have new coordinates given by

$$\Delta t' = \gamma \left(\Delta t - \frac{v}{c^2} \Delta x \right), \tag{23}$$

$$\Delta x' = \gamma(\Delta x - v\Delta t), \tag{24}$$

$$\Delta y' = \Delta y, \tag{25}$$

$$\Delta z' = \Delta z, \tag{26}$$

where the "boost" factor, represented by the Greek letter γ (pronounced "gamma"), is given by

[2] We can always rotate our coordinates so that the direction separating the events is along our choice of the x-direction.

$$\gamma = \frac{1}{\sqrt{1 - \frac{v^2}{c^2}}}. \tag{27}$$

In the example of Anna and Bob, we can have Anna using the coordinates (t', x', y', z') while Bob uses (t, x, y, z). In the new, moving, coordinate system we can calculate that Δs is given by

$$(\Delta s)^2 = c^2(\Delta t')^2 - (\Delta x')^2 - (\Delta y')^2 - (\Delta z')^2 \tag{28}$$

$$\begin{aligned}
&= c^2\gamma^2\left(-\tfrac{v}{c^2}\Delta x + \Delta t\right)^2 - \gamma^2(\Delta x - v\Delta t)^2 \\
&= \gamma^2\left(c^2\Delta t^2 + \tfrac{v^2}{c^2}\Delta x^2 - 2v\Delta x\Delta t\right) - \gamma^2(v^2\Delta t^2 + \Delta x^2 - 2v\Delta x\Delta t) \\
&= \gamma^2\left((c^2 - v^2)\,\Delta t^2 + \left(\tfrac{v^2}{c^2} - 1\right)\Delta x^2\right) \\
&= \gamma^2\left(1 - \tfrac{v^2}{c^2}\right)(c^2\Delta t^2 - \Delta x^2) \\
&= c^2\Delta t^2 - \Delta x^2,
\end{aligned} \tag{29}$$

which is precisely the same as we found in the old coordinates in Eq. (22). If the two events we were considering were the right distance apart so that a photon could be emitted from the first event and arrive at the second event, then the photon would have traveled a distance equal to the speed of light times Δt so that

$$\Delta x = c\Delta t \tag{30}$$

and also $\Delta s = 0$. In the boosted coordinate system Δs is also zero and if we calculate the speed of light in this boosted coordinate system it is given by distance traveled divided by the time interval:

$$\frac{\Delta x'}{\Delta t'} = \frac{\sqrt{c^2(\Delta t')^2 - (\Delta s)^2}}{\Delta t'} = \frac{\sqrt{c^2(\Delta t')^2}}{\Delta t'}, \tag{31}$$

$$= \frac{c\Delta t'}{\Delta t'} = c. \tag{32}$$

This is just what Einstein wanted; the speed of light measured in one coordinate system and in a boosted coordinate system has the same value: c. Or, in other words, the speed of light is independent of how fast we are moving when we measure the speed of light. The price of achieving this agreement with reality is that if we have simultaneous events in one coordinate system (so that $\Delta t = 0$) they are not simultaneous in another boosted coordinate system, since from Eq. (26) we have

$$\Delta t' = \gamma\left(-\frac{v}{c^2}\Delta x\right) \tag{33}$$

The minus sign tells us that, for positive v, the event with the larger value of x happens first in the new coordinate system.

We can also invert these relationships to go from Anna's coordinates to back to Bob's

$$\Delta x = \gamma(\Delta x' + \mathrm{v}\Delta t'), \tag{34}$$

$$\Delta t = \gamma\left(\Delta t' + \frac{\mathrm{v}}{c^2}\,\Delta x'\right) \tag{35}$$

Notice the change is essentially that the velocity in the formula goes from v to $-v$. This makes perfect sense, since if Bob sees Anna moving at velocity v in a certain direction, then Anna must see Bob moving at the same speed in the opposite direction. The minus sign in the velocity just tells us that v and $-v$ represent the same speed but in opposite directions.

In the example of Anna's simultaneously flashing lights discussed earlier, the boost velocity is in the easterly direction, since Bob sees Anna moving to the east, and v is positive. In this case we find that for Anna's simultaneous events ($\Delta t' = 0$), Bob sees a time difference of

$$\Delta t = \gamma \frac{v}{c^2}\,\Delta x', \tag{36}$$

and the event with the smaller value of x happens first in the boosted coordinates. This is just as we expect: the light leaving from the trailing hand (which covers more distance) meets up at Anna's nose at the same time as the light from the leading hand (which travels for a shorter time).

As the speed v gets closer and closer to the speed of light c, γ gets larger and larger, as shown in Fig. 13.1. If the velocity change is very small compared to the speed of light (as it is in everyday life) then the coordinate change is approximately

$$\Delta t' \approx \Delta t, \tag{37}$$

$$\Delta x' \approx (\Delta x - \mathrm{v}\Delta t), \tag{38}$$

$$\Delta y' = \Delta y, \tag{39}$$

$$\Delta z' = \Delta z, \tag{40}$$

which is indeed what we would naively expect: as the new coordinates move along the easterly (or x) direction with velocity v, points that were stationary in the old coordinates seem to move to the west with a speed[3] v.

Now consider an object that has length L_0 according to Anna who is moving along with it, how long will it look to Bob? Well, in order to measure its length Bob will look at the distance between the two ends at some fixed time according to his

[3] Or more technically with a velocity $-\mathrm{v}$ in the x direction, the minus sign tells us the motion is the direction opposite to increasing values of x.

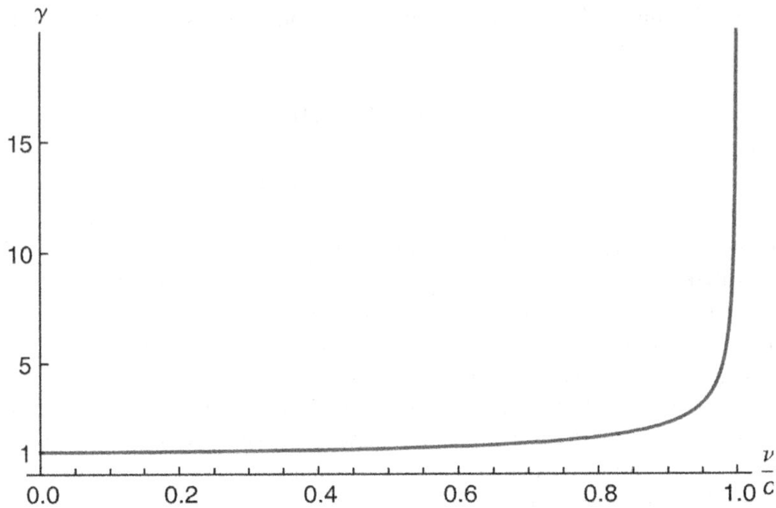

Fig. 13.1 γ stays very close to one for velocities much smaller than the speed of light but grows very large as the speed of light is approached

clock, which means we can use $\Delta t' = 0$ in Eq. (24). If we call the length that Bob measures L, then we must have

$$L_0 = \Delta x' = \gamma(\Delta x - v \cdot 0) = \gamma L. \tag{41}$$

So according to Bob, who sees the object moving at velocity v along the direction of its length, the length is actually

$$L = \frac{L_0}{\gamma}. \tag{42}$$

This is called relativistic length contraction. Objects appear to be contracting when they are moving relative to us. If it had been Bob holding the same object, Anna would have seen it as being shorter.

Similarly a time interval $\Delta t'$ between two events experienced by Anna at the same position in her coordinates will seem to be a longer time interval to Bob, who measures on his clock a time difference between the same two events given by Eq. (34):

$$\Delta t = \gamma \left(\Delta t' + \frac{v}{c^2} \cdot 0 \right) \tag{43}$$

$$\Delta t = \gamma \Delta t'. \tag{44}$$

The time Bob measures between the two events is longer than the time measured by the moving clock. It is always true that a clock traveling so that the two events are a fixed distance away measures a shorter time than any clock that is moving relative to the first; this is know as relativistic time dilation.

As we have seen what is simultaneous for Anna is not simultaneous for Bob, and what has a constant position for Anna is moving for Bob, so Bob's positions and times are a mixture of Anna's positions and times. If the position of an event measured by Bob is x, and the time of the event measured by Anna is t, and the corresponding position and time for Anna are x' and t', then there is a simple relation between the two sets of observations. The relation is even simpler if Anna agrees to set the zero of her x' and t' coordinates to agree with the zero of Bob's x and t coordinates. In this case the relation is

$$x' = \gamma(x - vt), \tag{45}$$

$$t' = \gamma\left(t - \frac{v}{c^2}x\right), \tag{46}$$

$$x = \gamma(x' + vt'), \tag{47}$$

$$t = \gamma\left(t' + \frac{v}{c^2}x'\right). \tag{48}$$

There is a similar set of formulae that relate velocities measured by Anna to those measured by Bob. Suppose that the velocity (that is the distance traveled divided by the time it takes to travel that distance) of a ball that Bob measures along the x-direction is

$$u_x = \frac{\Delta x}{\Delta t} \tag{49}$$

(this is a velocity parallel to Anna's direction of motion relative to Bob). Then Anna sees a velocity given by

$$u'_x = \frac{\Delta x'}{\Delta t'} \tag{50}$$

since again velocity is just distance traveled divided by the time it takes to travel that distance but now using Anna's coordinates. We can express this velocity in terms of the measurements Bob makes using Eqs. (23) and (24):

$$u'_x = \frac{\gamma\,(\Delta x - v\Delta t)}{\gamma\left(\Delta t - \frac{v}{c^2}\Delta x\right)}. \tag{51}$$

Canceling γ and dividing top and bottom by Δt we have

$$u'_x = \frac{\frac{\Delta x}{\Delta t} - v}{1 - \frac{\Delta x}{\Delta t}\frac{v}{c^2}}, \tag{52}$$

$$= \frac{u_x - v}{1 - \frac{u_x v}{c^2}}. \quad (53)$$

The numerator $u_x - v$ is the naive answer we would have expected before Einstein, while the denominator $1 - u_x\, v/c^2$ is the relativistic correction. Similarly, since $\Delta y' = \Delta y$ and $\Delta z' = \Delta z$ we find

$$u'_y = \frac{u_y}{\gamma\left(1 - \frac{u_x v}{c^2}\right)}, \quad (54)$$

$$u'_z = \frac{u_z}{\gamma\left(1 - \frac{u_x v}{c^2}\right)}. \quad (55)$$

If we want Bob's velocity in terms of Anna's velocity we would find

$$u_x = \frac{u'_x + v}{1 + \frac{u'_x v}{c^2}}. \quad (56)$$

Notice that if Anna is shining a beam of light forward and measures the velocity $u'_x = c$ then Bob will find

$$u_x = \frac{c + v}{1 + \frac{cv}{c^2}}, \quad (57)$$

$$= \frac{c(c + v)}{c(1 + \frac{v}{c})}, \quad (58)$$

$$= c, \quad (59)$$

so Bob will find that the speed of Anna's light beam is just the speed of light, independent of Anna's own velocity.

In Einstein's theory of special relativity when we examine distances and times, the invariant quantity under boosts is the difference of the squares of time component and the squares of spatial components. Similarly, for energy and momentum the invariant is the square of the energy (E) minus the squares of momentum components (p_x, p_y, p_z) times the speed of light squared:

$$E^2 - c^2(p_x^2 + p_y^2 + p_z^2). \quad (60)$$

Now if we are interested in a particular particle that happens to have a mass (m) we can always boost to a frame of reference (or coordinate system) where all the components of the momentum are zero (this is the frame of reference where the particle is at rest, aka the "rest frame"). Einstein also told us that the energy of a particle at rest is

$$E_{\text{rest}} = mc^2 \quad (61)$$

but this means that we know the value of the invariant above, since we know it in one particular frame, is invariant so it must be the same in any frame:

$$E^2 - c^2(p_x^2 + p_y^2 + p_z^2) = m^2c^4, \tag{62}$$

Solving for E we have

$$E = \sqrt{(p_x^2 + p_y^2 + p_z^2)c^2 + m^2c^4}. \tag{63}$$

We can always rotate our coordinates so that the x-direction is lined up with the direction the particles are moving, so that $p_y = 0$ and $p_z = 0$. Using

$$p = \gamma m v \tag{64}$$

we can find an even simpler expression for the energy:

$$\begin{aligned} E &= \sqrt{p^2c^2 + m^2c^4} = \sqrt{\tfrac{m^2v^2c^2}{1-v^2/c^2} + m^2c^4} \\ &= \sqrt{\tfrac{m^2v^2c^2+(1-v^2/c^2)m^2c^4}{1-v^2/c^2}} \\ &= \sqrt{\tfrac{m^2c^4}{1-v^2/c^2}} = \gamma mc^2 \end{aligned} \tag{65}$$

The kinetic energy (the energy associated with motion) is just the difference of total energy and the rest energy

$$KE = (\gamma - 1)mc^2. \tag{66}$$

For a velocity v much smaller than the speed of light, the square roots can be approximated using a Taylor series expansion (see Fig. 13.2)

$$\sqrt{1+x} \approx 1 + \frac{x}{2} - \frac{x^2}{8} + \cdots, \tag{67}$$

$$\frac{1}{\sqrt{1+x}} \approx 1 - \frac{x}{2} + \frac{3x^2}{8} + \cdots. \tag{68}$$

In general the Taylor series of a function $f(x)$ near $x = 0$ has the form

$$f(x) \approx f(0) + ax + \frac{b}{2}x^2 + \frac{c}{2\times 3}x^3 + \cdots \tag{69}$$

The value of a is fixed to be the slope (also called the first derivative) of the function at $x = 0$. If the slope is different on either side of $x = 0$ then the slope has its own slope and b is fixed to be the slope of the slope (or second derivative) and so on. If an approximation is needed around a point x_0 that is not equal to zero then we can write

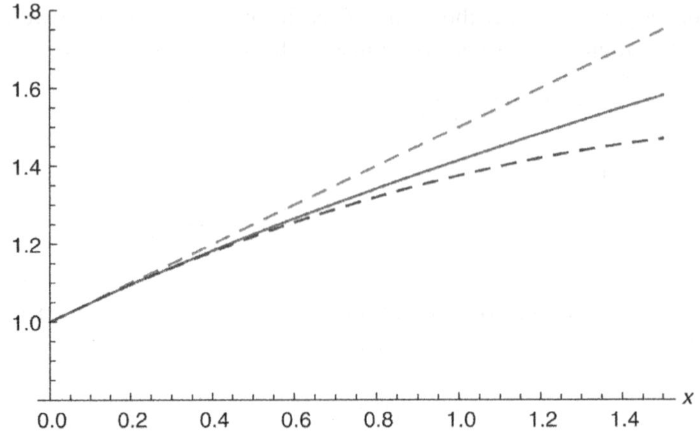

Fig. 13.2 The *solid line* shows $\sqrt{1+x}$, the *short dashes* show the first-order Taylor series approximation $1 + x/2$, and the *long dashes* show the second-order Taylor series approximation $1 + x/2 - x^2/8$

$$f(x) \approx f(x_0) + a(x - x_0) + \frac{b}{2}(x - x_0)^2 + \frac{c}{2 \times 3}(x - x_0)^3 + \cdots, \tag{70}$$

where a is fixed to be the slope of the function at $x = x_0$, b is the slope of the slope at $x = x_0$, and so on.

Using the Taylor series expansion (68) with $x = -\,v^2/c^2$ we find

$$KE \approx \frac{1}{2} m\,\mathrm{v}^2, \tag{71}$$

which is just what was used in physics before the advent of special relativity. This is just as it should be since for small velocities special relativity should reproduce what was known before Einstein.

13.3 Quantum Mechanics

The frequency, f, of oscillation of a light wave (in hertz) is related to the wavelength (the distance between neighboring peaks of the wave). Wavelength is usually represented by the Greek letter λ (pronounced "lambda"), and then the frequency is given by

$$f = \frac{c}{\lambda}, \tag{72}$$

where c is the speed of light. If the wavelength is given in meters, then the speed of light should be given in meters per second, so that meters cancel out, and the result has units of 1/second, which is equivalent to hertz. This formula actually works for any wave if c is replaced by the speed of the wave.

Planck's formula for the distribution of light in terms of the various possible wavelengths coming from a small window in a heated oven is known as the blackbody distribution. The distribution of light that comes from any heated object depends on how incoming light is reflected. The amount of light reflected is different for different frequencies and varies for different materials. To avoid this complication it is helpful to think about an idealized material that absorbs all light that falls on it. This idealized substance is called a blackbody. An oven with a very small window is a very good approximation to a blackbody since light falling into the window will bounce around inside the oven and has very little chance of being reflected back out of the window.

The formula that Planck found the light intensity as a function of wavelength for a blackbody at a temperature T (measured in kelvin or K for short) is

$$\rho(\lambda) = \frac{2ch}{\lambda^5(e^{hc/(\lambda k_B T)} - 1)}, \tag{73}$$

where h is Planck's constant (Planck originally called it the "Hilfsgröβe," which is German for "help-factor")

$$h = 6.63 \times 10^{-34}\,\mathrm{J\,s} = 4.14 \times 10^{-15}\mathrm{eV\,s} \tag{74}$$

and

$$k_\mathrm{B} = 8.62 \times 10^{-5}\mathrm{eV/K}, \tag{75}$$

is the Boltzmann constant which was already known from the study of thermodynamics. One often sees Planck's constant written in another form

$$\hbar = \frac{h}{2\pi} = 1.06 \times 10^{-34}\,\mathrm{J\,s} = 6.58 \times 10^{-16}\mathrm{eV\,s}. \tag{76}$$

Planck's constant is very small compared to ordinary day-to-day quantities; it has units of energy times seconds and 1 J is about the energy a cup has when it is knocked off a counter and hits the floor. An eV (short for electronvolt) is the energy that a single electron gets when it is accelerated through 1 V difference in an electric field. (A typical battery provides 1.5 V.) Even though 1 eV is a typical atomic energy scale, Planck's constant is still tiny in eV seconds since atomic timescales are much smaller than 1 s. Since Planck's constant is so small we can consider the approximation that results from taking h to be zero; this gives

$$\rho_c(\lambda) = \frac{k_\mathrm{B}T}{\lambda^4} \tag{77}$$

which was indeed the standard prediction for a blackbody spectrum before the development of quantum mechanics. This formula is a good approximation to Planck's formula for long wavelengths, that is when

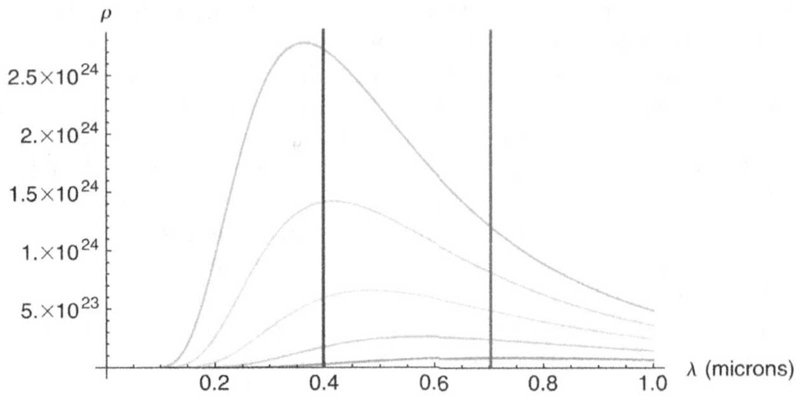

Fig. 13.3 The distribution of electromagnetic waves in a hot cavity versus the wavelength. The *curves* correspond to different temperatures starting with 8000 K at the top going down to 4000 K in steps of 1000 K. The *vertical line* near 0.4 μm represents the shortest wavelength in the visible spectrum of light (*violet*) while the *vertical line* on the *right* represents the longest visible wavelength (*red*). A cavity heated to 1000 K has an *orange* glow, when it is heated to 8000 K it is blue-hot

$$\lambda \gg \frac{hc}{k_B T}. \tag{78}$$

The best measured blackbody distribution ever measured is that of the cosmic microwave background radiation. This is the radiation that is left over in the universe from the hot Big Bang. The universe has expanded and cooled since that time. Careful measurements by the COBE satellite showed that this microwave background radiation agrees perfectly with Planck's formula with a temperature[4] of 2.7 K. The radiation is called microwave because Planck's formula shows that the peak of curve at this very low temperature is for wavelengths around 1000 μm (also known as a millimeter) which is in the microwave range.

Einstein concluded from his study of the photoelectric effect that the energy of a photon is related to its frequency by Planck's constant:

$$E = hf = \hbar\omega. \tag{79}$$

Here the Greek letter ω (written as "omega," pronounced as "oh-may-guh") represents an "angular frequency" as we will see later. There is also a corresponding relation between momentum and wavelength

$$p = \frac{h}{\lambda} = \hbar k \tag{80}$$

[4]The boiling point of water is 473 K and the freezing point of water is 373 K, so 2.7 K is *very* cold.

which agrees with the formula for energy (63) once we remember that a photon has a mass $m = 0$. The wavenumber k is introduced in analogy to ω and is just another way of specifying the wavelength

$$k = \frac{2\pi}{\lambda}. \tag{81}$$

Einstein showed that there should be a maximum kinetic energy for electrons kicked out of a metal by the photoelectric effect:

$$KE_{\max} = hf - W, \tag{82}$$

which tells us that there is some minimum energy, W, required to removed the electron (if the electron is deeper in the metal, then the amount of energy required to remove it goes up) and that the electron absorbs an amount of energy from the photon equal to hf. If light is really made of particles, then you might expect to be able to scatter a photon and an electron just like scattering two electrons or even two billiard balls. The scattering of photons and electrons was first observed by Arthur Holly Compton in 1923, which earned him the Nobel Prize 4 years later. One can calculate the final energy that the photon has after scattering off an electron by using the conservation of energy and momentum just as one would do for billiard ball scattering except here our colliding particles are relativistic (one is even massless) and we can't use the non-relativistic formula for kinetic energy, as in Eq. (71), that we would have used for billiard balls. For the photon we start out with an energy

$$E_\gamma = \frac{hc}{\lambda} \tag{83}$$

and the electron starts at rest with energy

$$E_e = m_e c^2, \tag{84}$$

where m_e is the mass of an electron:

$$m_\mathrm{e} = 9.11 \times 10^{-31}\,\mathrm{kg} = 511\,\mathrm{keV/c^2}. \tag{85}$$

Again we see that everyday units like kilograms (or pounds) are too large and clumsy for delicate elementary particles. We can partially get around this by using the corresponding rest energy of the electron, measured in kilo electron volts (keV), and quoting the mass as this energy divided by the speed of light squared.

If the photon emerges from the scattering at an angle θ from its initial direction and with wavelength λ', then the shift in the wavelength is

$$\lambda' - \lambda = \frac{h}{m_\mathrm{e} c}(1 - \cos\theta). \tag{86}$$

So the final photon energy is

$$E' = hf' = \frac{hc}{\lambda'}. \tag{87}$$

If the photon goes straight through, which means $\theta = 0$ so $\cos\theta = 1$ there is no change in wavelength, so in this case the photon's energy does not change. For other angles, the final wavelength is longer, so the photon has lost energy to the electron. When the photon bounces straight back, then $\cos\theta = -1$, and the photon suffers maximum energy loss.

13.4 Traditional Extra Dimensions

A standing wave on a string that goes between $x = 0$ and $x = L$ can be described by a simple mathematical formula. We will use the Greek letter ψ (written as "psi" but pronounced as "sigh"), to represent the height of the string above its resting height as a function of position and time

$$\psi(x,t) = A\sin(kx + \phi)\cos(\omega(t - t_0)), \tag{88}$$

where A is the maximum size of the wave, while k and ω are determined by the wavelength and frequency:

$$k = \frac{2\pi}{\lambda}, \quad \omega = 2\pi f. \tag{89}$$

The factors of 2π are introduced for convenience in working with the trigonometric functions, since they are periodic with period 2π, which means $\sin(\theta) = \sin(\theta + 2n\pi)$, for any θ and any integer value for n. Since we want a wave that doesn't move at the endpoints we can take $\phi = 0$, since $\sin(k0) = 0$. (If we had wanted a wave that was free to wiggle at $x = 0$ we could choose $\phi = \pm\pi/2$.) Now if our standing wave doesn't wiggle at the other end where $x = L$, then we need

$$\sin(kL) = 0, \tag{90}$$

so

$$kL = \frac{2\pi L}{\lambda} = n\pi \tag{91}$$

which means

$$\lambda_n = \frac{2L}{n}, \tag{92}$$

where n can take on integer values $n = 1, 2, 3, \ldots$, and we have introduced a subscript n on the wavelength λ in order to keep track of which wave we are talking about. It doesn't make sense to use $n = 0$ since then there is no wave: the function $\psi(x,t)$ is always zero for $n = 0$. These kinds of waves are shown in Fig. 12.14 for $n = 1, 2, 3, 4$. The frequencies associated with these standing waves are

$$f = \frac{c}{\lambda_n} = \frac{n\,c}{2L}. \tag{93}$$

For a wave that is free to wiggle on both ends, let's take $\phi = \pi/2$, then we find the wavelength from

$$\sin(kL + \phi) = \pm 1, \tag{94}$$

so

$$kL + \phi = \frac{2\pi L}{\lambda} + \phi = n\pi + \frac{\pi}{2} \tag{95}$$

and we find

$$f = \frac{c}{\lambda_n} = \frac{n\,c}{2L}, \tag{96}$$

where now we can take $n = 0, 1, 2, 3, \ldots$ since for $n = 0$ the function $\psi(x,t)$ is just a non-zero constant function. This kind of wave is shown on the left-hand side of Fig. 12.28 for $n = 0, 1, 2 \ldots$.

The third possible case is a wave that is free to wiggle at end and can't wiggle at the other end. If we take the non-wiggling end to be at $x = L$ then $\phi = 0$ and in order to have

$$\sin(kL) = \pm 1 \tag{97}$$

we need

$$kL = \frac{2\pi L}{\lambda} = n\pi + \frac{\pi}{2} \tag{98}$$

so we find

$$f = \frac{c}{\lambda_n} = \frac{c}{L}\left(\frac{n}{2} + \frac{1}{4}\right). \tag{99}$$

where again we can take $n = 0, 1, 2, 3, \ldots$ but for $n = 0$ the function $\psi(x, t)$ is not a constant function. This kind of wave is shown on the right-hand side of Fig. 12.28 for $n = 0, 1, 2$.

If we generalize the energy formula (63) for a particle in 4 space–time dimensions, to a particle that can also move in an extra dimension then we can write

$$E = \sqrt{(p_x^2 + p_y^2 + p_z^2 + p_5^2)c^2 + m^2c^4}, \tag{100}$$

where p_5 is the component of momentum along the extra dimension. Comparing with usual formula (63) we see that we have what looks like an effective mass

$$m_{\text{effective}}^2 c^4 = p_5^2c^2 + m^2c^4. \tag{101}$$

If we are interested in the case of a 5-dimensional particle with $m = 0$ (as would be the case for the photon), we find

$$m_{\text{effective}} = \frac{p_5}{c}. \tag{102}$$

We can use the relation between the momentum in the extra dimension and its wavelength, and write the effective mass in terms of the size L of the extra dimension if we know whether the standing wave is free or not to wiggle at the ends. With a standing wave going to zero at the endpoints then we have

$$m_{\text{effective}} = \frac{h}{c\lambda_n} = \frac{nh}{2cL}, \quad n = 1,2,3,\ldots. \tag{103}$$

Here the lightest state has mass

$$m_1 = \frac{h}{2cL}. \tag{104}$$

With the wave free to wiggle at the endpoints we have

$$m_{\text{effective}} = \frac{h}{c\lambda_n} = \frac{nh}{cL}, \quad n = 0,1,2,3,\ldots. \tag{105}$$

Here the lightest state is massless. With a wave free to wiggle at one end and fixed not to wiggle at the other end we have

$$m_{\text{effective}} = \frac{h}{c\lambda_n} = \frac{h}{cL}\left(\frac{n}{2} + \frac{1}{4}\right), \quad n = 0,1,2,3,\ldots. \tag{106}$$

Here the lightest state has mass

$$m_0 = \frac{h}{4cL}. \tag{107}$$

In the original Kaluza–Klein theory the extra dimension is curled up in a circle, so the previous discussion has to be slightly modified. If we wrap up the interval from 0 to L as a circle then when we get $x = L$ we are really back to $x = 0$. This means we cannot have the case where the wave is zero at one end and wiggling at the other end. So the tower of Kaluza–Klein particles has masses given by Eq. (103).

13.5 Einstein's Gravity

Recall that in special relativity the invariant separation between two points in space–time is

$$(\Delta s)^2 = c^2(\Delta t)^2 - (\Delta x)^2 - (\Delta y)^2 - (\Delta z)^2. \tag{108}$$

When we need to include gravity we have to generalize this a little. First the gravitational field can vary from point to point, so in order to write a simple formula we need to look at arbitrarily small (or infinitesimal) separations. Then if we wish can add up the invariant separations for all the segments along some path to get the total separation between two points at a finite distance (that is we can integrate to get the total separation). So instead of writing Δx (a finite difference in the x-direction) we will write things in terms of dx, the infinitesimal difference in the x-direction. So in the absence of gravity we can write an infinitesimal separation:

$$ds^2 = c^2(dt)^2 - (dx)^2 - (dy)^2 - (dz)^2. \tag{109}$$

In the presence of a gravitational field coefficient functions can appear in the invariant separation:

$$ds^2 = f(x,t)c^2dt^2 - g(x,t)(dx^2 + dy^2 + dz^2). \tag{110}$$

This type of equation and the functions that appear in it are referred to in general as the *metric* of the space, since it tells how distances are measured between two points in the space. In a completely general gravitational field there can even be terms proportional to $dx\ dy$ or $dx\ dt$. The amount that $f(x, t)$ and $g(x, t)$ differ from one tells us the strength of the gravitational field. Of course Einstein told us that the strength of the gravitational field depends on our choice of a coordinate system. If we let ourselves fall in the gravitational field, then, in terms of our personal point-of-view coordinate system that is moving with us, the separation (110) reduces to Eq. (109) for sufficiently small distances and times.

For problems with rotational symmetry it is often easier to use spherical coordinates rather than the (x, y, z) (aka Cartesian coordinates) that we have been using. For example, if we are interested in the gravitational field around a uniform spherical mass (the Earth, for example, is almost spherical) then the force will only depend on the distance we are from the center of the spherical mass not on the direction in space. We can represent the distance (aka the radial distance) from the center of the mass by r where

$$r = \sqrt{x^2 + y^2 + z^2}. \tag{111}$$

We can also consider the case where the distribution of mass varies inside the massive body. If the density itself is only a function of r then again the gravitational force will not depend on direction only on r. Of course if we want to fully specify our position in space we need more information than just r, but we can specify it in terms of two angles, which are traditionally referred to by the Greek letters θ and ϕ (ϕ is written as "phi" and pronounced as "fie"). If we draw a line from the center of our coordinate system (where $x = y = z = 0$) to our position a distance r away, then the angle between the z-direction and the line to r is θ. This means that θ can be between 0° and 180°. On the surface of the Earth lines of constant θ correspond to the lines of latitude, although the degrees of latitude are measured from the equator at 0° up to 90° north (90° north being at the North Pole) and similarly in the southern hemisphere, from 0° to 90° south. The angle ϕ represents the angle between the line to our position and the plane made by the x- and z- directions, so ϕ can be between 0° and 360°. On the surface of the Earth lines of constant ϕ are called lines of longitude. A little trigonometry reveals the relation between these angles and the Cartesian coordinates:

$$\begin{aligned} x &= r \sin\theta \cos\phi, \\ y &= r \sin\theta \sin\phi, \\ z &= r \cos\theta \end{aligned} \tag{112}$$

and

$$ds^2 = c^2 dt^2 - dr^2 - r^2 \, d\Omega_2, \tag{113}$$

where

$$d\Omega_2 = d\theta^2 + \sin^2\theta d\phi^2 \tag{114}$$

is the metric on a sphere of radius 1. The factor of $\sin^2\theta$ in front of $d\phi^2$ tells us that a trip around the world following the line of latitude 70° north (so $\theta = 90 - 70 = 20°$) is much shorter than going around the world following the equator (where $\theta = 90°$). In fact the Northern trip is only about

$$\sin\left(20°\right) \approx \frac{1}{3} \tag{115}$$

or one-third as far.

A particularly interesting application of Einstein's general relativity is the solution of the invariant separation (aka the line element or metric) around a black hole of mass M, which can be written as

$$ds^2 = f(r)\, c^2 dt^2 - \frac{1}{f(r)} dr^2 - r^2 \, d\Omega_2, \tag{116}$$

where

$$f(r) = 1 - \frac{2 G_N M}{r c^2}, \tag{117}$$

and G_N is Newton's gravitational constant

$$G_N = 6.67 \times 10^{-11} \frac{\text{m}^3}{\text{kg s}^2} \tag{118}$$

which appears in Newton's universal law of gravitation. This latter formula, which gives the strength of the force between two masses, m_1 and m_2, separated by a distance R, is

$$F = \frac{G_N m_1 m_2}{R^2}. \tag{119}$$

The gravitational force is always directed along the line between the two masses and acts to pull the masses toward each other.

We can see that something interesting happens when the value of radial coordinate r in Eq. (116) goes through the value

$$r_H(M) = \frac{2G_N M}{c^2}. \tag{120}$$

Imagine falling into a black hole; you start some distance $r > r_H(M)$ away from the center. As you get closer to r_{H}, the metric factor $f(r)$ is getting smaller, at $r = r_H(M), f(r_H(M))$ is zero, and as your position r gets even smaller the metric factor $f(r)$ becomes negative. This means that inside the radius r, r behaves as a time direction and t behaves as a space direction. This also means that you can't go back outside of the black hole, unless you can increase your value of r, which would require moving backward in the time direction. Since no one knows how to move backward in time it seems that once you fall inside the sphere at radius $r_H(M)$ you cannot get back out. This sphere is called the black hole horizon.

Metric (116) is useful for more than just black holes, it can be applied outside of any spherical mass distribution as long as its physical size r_S is larger than the horizon size $r_H(M)$ for its mass. If $r_S > r_H(M)$ then we can use metric (116) for $r > r_S$ while inside r_S there is a more complicated situation. Given the mass of the Earth we can find the corresponding horizon size:

$$M_{\text{E}} = 5.97 \times 10^{24}\ \text{kg}, \tag{121}$$

$$r_{\text{H}}(M_{\text{E}}) = 8.8\ \text{mm}, \tag{122}$$

which is certainly much less than the radius of the Earth

$$r_{\text{E}} = 6{,}370\ \text{km} = 3{,}960\ \text{miles}. \tag{123}$$

If one could crush a mass as big as the Earth inside a sphere of radius 8.8 mm then you would have formed a black hole. Suppose we start falling toward the Earth from a position r_0 which is above the surface of the Earth, so $r_0 > r_E$. For a very short time interval Einstein tells us that in a small region of space a freely falling person feels no gravitational field. Such a person would use a metric similar to (113) but we will put a $(\hat{\ })$ on the coordinates to remind us that these are the coordinates used by the falling person. We need to find the relation between the falling coordinates $\hat{r}$, $\hat{t}$ and the coordinates in metric (116) r, t. The simplest possible relation is that near r_0 we have

$$r = r_0 + b\,\hat{r} \tag{124}$$

and that the time coordinate is rescaled as well:

$$t = e\,\hat{t}, \tag{125}$$

where b and e are constants. Now we can write

$$dr = b\,d\hat{r}, \tag{126}$$

$$dt = e\,d\hat{t}. \tag{127}$$

Near r_0 we can approximate the metric function $f(r)$ by the first terms in a Taylor series expansion (see Fig. 13.4)

$$\frac{1}{1+x} \approx 1 - x + x^2 - x^3 + \ldots, \tag{128}$$

which gives

$$f(r) \approx 1 - \frac{2G_N M_E}{r_0\,c^2} + \frac{2G_N M_E(r - r_0)}{r_0^2\,c^2}. \tag{129}$$

So to lowest order in $\hat{r}$ and $\hat{t}$ the metric becomes

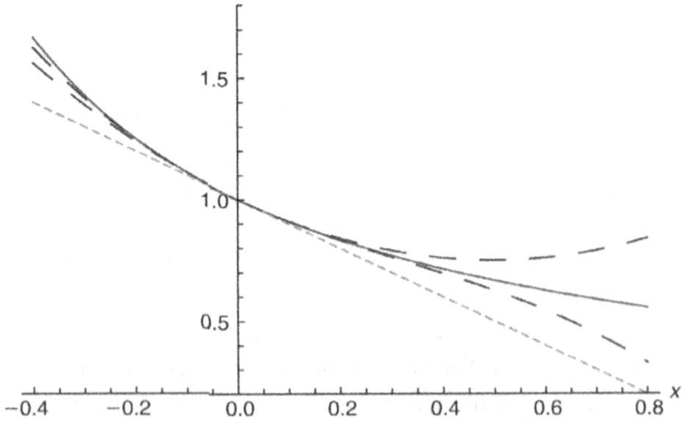

Fig. 13.4 The *solid line* shows $1/(1 + x)$, the *short dashes* show the first-order Taylor series approximation $1\text{-}x$, the *medium dashes* show the second-order Taylor series approximation $1\text{-}x + x^2$, and the *long dashes* show the third-order Taylor series approximation $1\text{-}x + x^2\text{-}x^3$

$$ds^2 = \left(1 - \frac{2G_N M_E}{r_0\, c^2}\right) c^2 e^2\, d\hat{t}^2 \tag{130}$$

$$- \left(1 + \frac{2G_N M_E}{r_0\, c^2}\right) b^2\, d\hat{r}^2 - r_0^2 d\Omega_2, \tag{131}$$

and with

$$b = 1 - \frac{G_N M_E}{r_0\, c^2}, \tag{132}$$

$$e = 1 + \frac{G_N M_E}{r_0\, c^2} \tag{133}$$

we find

$$ds^2 = c^2\, d\hat{t}^2 - d\hat{r}^2 - r_0^2 d\Omega_2, \tag{134}$$

What Eq. (132) tells us is that a falling person and a person at a constant value of r (which means they must be firing some rockets continuously to fight the gravitational field that is trying to pull them down) see the same object appear to have different lengths. An object measured by the falling person to have length L along the direction he is falling is smaller according to the stationary person who measures a length

$$b\, L = \left(1 - \frac{G_N M_E}{r_0\, c^2}\right) L. \tag{135}$$

So for a light beam falling toward the Earth the wavelength gets progressively shorter as it falls, and we say there is a gravitational blueshift. Conversely a light beam rising from the Earth gets progressively longer; so it experiences a gravitational redshift. Similarly a time interval T between two events experienced by the falling person is longer according to the stationary observer who records a time interval

$$eT = \left(1 + \frac{G_N M_E}{r_0\, c^2}\right) T. \tag{136}$$

So a clock that is closer to the Earth runs more slowly than a clock that is farther from the Earth. This is called a gravitational time dilation.

In the Kaluza–Klein theory there are extra terms in the metric

$$\begin{aligned} ds^2 = & f(x,t)c^2 dt^2 - g(x,t)(dx^2 + dy^2 + dz^2) - \rho(x,t)dx_5^2 \\ & - A_0(x,t)dtdx_5 - A_1(x,t)dxdx_5 \\ & - A_2(x,t)dydx_5 - A_3(x,t)dzdx_5, \end{aligned} \tag{137}$$

where x_5 is the coordinate along the extra dimension.

13.6 The Theory Formerly Known as String

If the idea of quarks connected by color flux tubes making up most to the mass of your body (since protons and neutrons are about 2000 times heavier than electrons) seems obscure, then perhaps Murray Gell-Mann's choice of the name quark was appropriate. The name came from the famous novel "Finnegans Wake" by James Joyce. "Finnegans Wake" is partly famous for being a famous novel that almost no one has read. The line from the book is "Three quarks for Muster Mark!" and although Joyce intended quark to rhyme with Mark, physicists pronounce it so that it rhymes with cork. When Gell-Mann introduced the idea of quarks, he treated them as a convenient mathematical fiction that made it easier to understand the zoo of strongly interacting particles that were being found at that time. Later when string theory was first being applied to the strong interactions, many people were excited that the real underlying elements had been found. As it turned out quarks are real, and the strings, at least in the case of strong interactions, were only convenient mathematical approximations to the complicated interactions of quarks and gluons. Current string theorists hope that strings (and branes) are the real underlying elements behind Einstein's gravity ... only time will tell.

Returning to gravity, in a 4-dimensional space–time the force between two small masses follows Newton's universal law of gravitation (119), but in a D-dimensional space–time we would have

$$F = \frac{\kappa_D^2 m_1 m_2}{R^{(D-2)}}. \tag{138}$$

Essentially the force of gravity decreases with the area of the hypersphere that the gravitational field lines have to be spread over. In a 3-dimensional space the force lines are spread over an ordinary sphere, and so the force falls like one over the distance squared. The intensity of light falls off the same way for essentially the same reason. Now a force is equivalent to mass times acceleration which has units of mass times length per second squared, and energy is equal to mass times the speed of light squared so it has units of mass times length squared per second squared. So we see that force has units of energy over length, which means that for $D = 4$, $G_N = \kappa_4^2$ has dimensions of energy times length over mass squared. In fact we can define a Planck mass scale (as Planck indeed did) just using the three basic constants G_N, $\hbar$, and c:

$$G_N = \kappa_4^2 = \frac{\hbar c}{M_{\text{Planck}}^2}, \tag{139}$$

so we find

$$M_{\text{Planck}} = 1.2 \times 10^{19}\text{GeV}/c^2. \tag{140}$$

In a quantum theory a wave with an energy equal to M_{Planck} has a wavelength equal to the Planck length:

$$L_{\text{Planck}} = \frac{\hbar c}{M_{\text{Planck}}} = 1.6 \times 10^{-33}\ \text{cm}. \tag{141}$$

If we now turn to a universe with D space – time dimensions we have that the D-dimensional Newton's constant κ_D^2 has dimensions of energy times $D-3$ powers of length over mass squared. So in such a world we would relate the D-dimensional Newton's constant to a Planck mass by

$$\kappa_D^2 = \frac{c^3}{\hbar}\left(\frac{\hbar}{c\,M_{\text{Planck}}}\right)^{D-2}. \tag{142}$$

Specializing to 11-dimensional M-theory we have that the 11-dimensional Newton's constant κ_{11}^2 has dimensions of energy times eight powers of length over mass squared. So in this 11-dimensional universe we relate the 11-dimensional Newton's constant to a Planck mass by

$$\kappa_{11}^2 = \frac{c^3}{\hbar}\left(\frac{\hbar}{c\,M_{\text{Planck}}}\right)^{9} = \frac{\hbar^8}{c^6\,M_{\text{Planck}}^9}. \tag{143}$$

Now let us consider the energy per unit volume (aka the *tension*) of various branes which are given in terms of the Planck mass. In M-theory there are only two fundamental type of branes: 2-brane and 5-brane. The 2-brane is a brane which fills 2 spatial dimensions (which is just like an ordinary membrane), while the 5-brane fills five spatial dimensions. In M-theory the energy per unit area (or tension) of the 2-brane is

$$T_2 = M_{\text{Planck}}^3 \frac{c^4}{\hbar^2}. \tag{144}$$

Similarly the 5-brane of M-theory has tension equal to

$$T_5 = M_{\text{Planck}}^6 \frac{c^7}{\hbar^5}. \tag{145}$$

Now if we curl up one of the dimensions of M-theory (just like Kaluza and Klein proposed so long ago) then we get a universe with 10 space – time dimensions which turns out to be described by type IIA string theory. This is one of the examples where Witten was able to show that the original string theories were special cases of M-theory. Now type IIA string theory has four types of branes: 1-brane, 2-brane, 4-brane, and 5-brane. The 2-brane and 5-brane of type IIA string theory have same tensions as in 11-dimensional theory since they are just the same branes as before. We get two new types of branes because the branes can get wrapped around the curled-up dimension. If R_{10} is the length of the curled-up dimension then the

1-brane (which looks just like a string on length scales that are large compared to R_{10}) has a tension (or energy per unit length) given by

$$T_1 = R_{10}\, M_{\text{Planck}}^3 \frac{c^4}{\hbar^2}, \tag{146}$$

while the 4-brane tension is

$$T_4 = R_{10}\, M_{\text{Planck}}^6 \frac{c^7}{\hbar^5}. \tag{147}$$

It turns out that the 1-brane formed this way is just the "fundamental" string of Type IIA string theory. In type IIA string theory the tension of the "fundamental" string is written in terms of the "string mass," m_s:

$$T_1 = R_{10}\, M_{\text{Planck}}^3 \frac{c^4}{\hbar^2} \equiv \frac{m_s^2 c^3}{\hbar}. \tag{148}$$

We can also express the tensions of the other branes in terms of m_s and the dimensionless type IIA string coupling:

$$g_s = \left(\frac{R_{10}\, M_{\text{Planck}}\, c}{\hbar} \right)^{3/2}. \tag{149}$$

One finds

$$T_2 = \frac{m_s^3}{g_s} \frac{c^4}{\hbar^2}, \quad T_4 = \frac{m_s^5}{g_s} \frac{c^6}{\hbar^4}, \quad T_5 = \frac{m_s^6}{g_s^2} \frac{c^7}{\hbar^5}. \tag{150}$$

Since branes are a type of nonperturbative *soliton*,[5] a mathematician would not be surprised to see inverse powers of the coupling squared, g_s^2, appearing in T_5. But the inverse of g_s in the tensions of the 2-brane and 4-brane is unusual. This is actually a universal feature of a special type of branes called D-branes, where the D is short for Dirichlet. They are named in honor of the mathematician Dirichlet since the strings that end on these branes have Dirichlet boundary conditions. When we looked at standing waves on a string we used Dirichlet boundary conditions when the ends were not free to wiggle.

It was the study of 3-dimensional D-branes in type IIB string theory that led Maldacena to his famous conjecture relating a 4-dimensional gluon-like theory to gravity in a 5-dimensional anti-de Sitter space. In order to understand a little more about the conjecture we will need to know a little more about spheres, anti-de Sitter spaces, and supersymmetric gauge theories.

An ordinary sphere has a 2-dimensional surface (so mathematicians like to call it a 2-sphere, or S^2 for short) and can live in a space with 3 dimensions. If we rotate a

[5] A soliton is a special type of single wave pulse that can travel forever without breaking apart.

perfect sphere around its center nothing seems to change. We can perform the twist along any direction in the 3 spatial dimensions, that is we can choose any rotation out of the set of all possible rotations in 3 dimensions. Because this set of rotations has some special properties, mathematicians call it a group, or more precisely a rotation group, they have even given this group the name:[6] $SO(3)$. The surface of a 2-sphere with radius R can be found by solving the equation

$$R^2 = x^2 + y^2 + z^2, \tag{151}$$

just as the circle, or 1-sphere, is described by

$$R^2 = x^2 + y^2. \tag{152}$$

Similarly, a sphere with a 5-dimensional surface, aka a 5-sphere or S^5, can live in a 6-dimensional space and is unchanged by $SO(6)$ rotations in this space. The metric for a flat 6-dimensional space is

$$ds^2 = -(dY_1^2 + dY_2^2 + dY_3^2 + dY_4^2 + dY_5^2 + dY_6^2). \tag{153}$$

The hyper-surface of a 5-sphere with radius R in this space is found by solving the equation

$$R^2 = Y_1^2 + Y_2^2 + Y_3^2 + Y_4^2 + Y_5^2 + Y_6^2. \tag{154}$$

Spheres are spaces with constant positive curvature, so the sum of interior angles of a triangle add up to more than 180° (see the left diagram in Fig. 12.18).

A 5-dimensional anti-de Sitter space, aka AdS_5, can be embedded in 6-dimensional space with a metric:

$$ds^2 = dX_0^2 + dX_5^2 - (dX_1^2 + dX_2^2 + dX_3^2 + dX_4^2). \tag{155}$$

The hyper-surface corresponding to the anti-de Sitter space can be found by solving

$$R^2 = X_0^2 + X_5^2 - \left(X_1^2 + X_2^2 + X_3^2 + X_4^2\right), \tag{156}$$

which describes a generalization of an hyperboloid as shown in Fig. 13.5.

Five-dimensional anti-de Sitter space is a space with a constant negative curvature and a negative cosmological constant. Anti-de Sitter space is also a space with constant negative curvature, so the sum of interior angles of a triangle adds up to less than 180° (see the right diagram in Fig. 12.18). The group of transformations that leaves 5-dimensional anti-de Sitter space unchanged is called $SO(4, 2)$. The distinction 4, 2 rather than 6 (which is 4 + 2) refers to the fact that in Eq. (155) four

[6]The name stands for special orthogonal transformations in 3 dimensions.

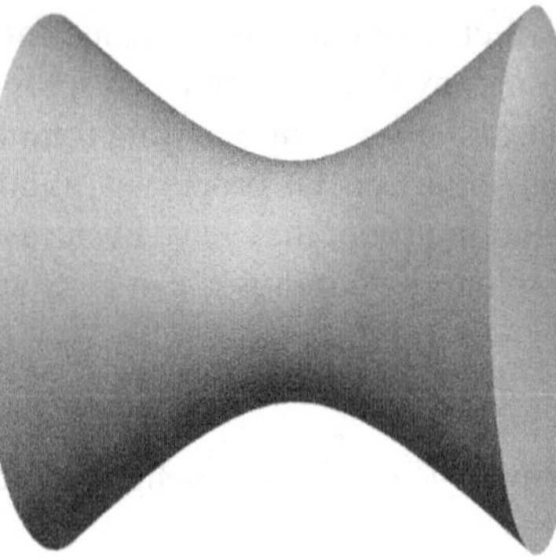

Fig. 13.5 Anti-de Sitter space as an hyperboloid embedded in a higher dimensional space.

of the terms (corresponding to the spatial directions) have a minus sign in front of them while two of the terms (corresponding to the time directions) have a positive sign in front of them.

Returning to Maldacena, he was studying a theory with N copies of 3-dimensional D-branes all stacked on top of one another inside a 10-dimensional space – time. Each 3-dimensional brane fills up a 3-dimensional space and moves through time, or in other words, it would look like something we would call a universe. Strings ending on a D-brane at each end can connect the N different D-branes in N^2 different ways. These strings can act like N^2 different gluons, and when the D-branes have zero separation between them (that is they are all in exactly the same point in the extra dimensions), then the strings have zero length, and thus can be created with zero energy like a massless particle. A theory with massless particles like gluons or photons is called a gauge theory. So Maldacena's toy universe describes a gauge theory, but it is a special type of gauge theory: it is also supersymmetric and in fact has four supersymmetries.

Supersymmetry is a symmetry that mixes bosons and fermions. This is quite unusual because bosons and fermions are two very different types of particles. Two identical fermions cannot be in the same place at the same time, while identical bosons prefer to be in the same place at the same time: the more the merrier. Electrons and quarks are fermions, and it is the electrons in your body and every other material object that make them feel solid when they are really mostly empty space. When two objects come in contact, when the rubber meets the road, or a bat hits a ball, the electrons in the ball cannot be in the same region as any other electron, and because of quantum mechanics any electron does not exist at just a point but is spread out in space like a wave. Photons, gluons, W's, and Z's (all the force carriers), on the other hand, are bosons. Lasers can generate such intense light beams because you can put many identical photons, traveling exactly the same way, at the same region at the same time. The idea of force carrier bosons that can interact with themselves (like gluons do) originated with Yang and Mills, so this gluon-like supersymmetric theory has the technical name $\mathcal{N} = 4$ supersymmetric Yang – Mills—quite a mouthful. For each gluon in this theory there are four "superpartner" fermions (called gauginos); this is a consequence of having four supersymmetries, with one supersymmetry there would only be one gaugino for each gluon. For each gluon there are also six other bosons that resemble the Higgs boson more than anything else; we'll call them *scalars* for short. The presence of six scalars is also a consequence of having four supersymmetries, with just one supersymmetry a gluon would not have a boson superpartner at all, while for two supersymmetries there are two scalars for each gluon. The six scalars in the $\mathcal{N} = 4$ theory are completely equivalent to each other, so there is a symmetry that can mix them up, and that symmetry, it turns out not surprisingly, is $SO(6)$. It also turns out that a 4-dimensional theory that is scale invariant, that is it looks the same at all length scales and only has gluons, gauginos, and scalars has one more symmetry (technically called a conformal symmetry) and the group of these symmetry transformations is, surprisingly, $SO(4, 2)$. So $\mathcal{N} = 4$ supersymmetric Yang–Mills is a particular example of a conformal field theory,[7] or CFT for short. Finally we see that the 4-dimensional, $\mathcal{N} = 4$ supersymmetric Yang–Mills theory has the same symmetries $SO(6)$ and $SO(4, 2)$ as a 5-sphere and a 5-dimensional anti-de Sitter space.

Alternatively, we can directly study the low-energy effective theory of this stack of N D-branes. The D-branes are a source for gravity just like any other mass or energy, and placing N of them together curves the 10-dimensional space that they live in. Calling the coordinates that span a 3-dimensional D-brane x, y, z, and the distance from the branes r, Maldacena found that the metric around the D-branes is

$$ds^2 = f^{-1/2}\left(-dt^2 + dx^2 + dy^2 + dz^2\right) + f^{1/2}\left(dr^2 + r^2 d\Omega_5^2\right), \tag{157}$$

where

$$f = 1 + \left(\frac{R}{r}\right)^4, \tag{158}$$

R is a curvature radius

$$R^4 = \frac{g_s N \hbar^2}{4\pi m_s^4 c^6}, \tag{159}$$

and $d\Omega_5^2$ represents the metric on the surface of a 5-sphere.

Since the coefficient of dt^2 in Eq. (157) depends on r, a person at position r who measures an energy E_r sees a red-shifted energy relative to an observer at $r = \infty$. The observer at $r = \infty$ would measure the energy of the same state to be

$$E = f^{-1/4} E_r. \tag{160}$$

[7] Quantum field theory is the name given to the tools we used to calculate how particles behave while taking into account both quantum mechanics and special relativity.

Taking the low-energy limit corresponds to considering states close to the D-branes with $r \to 0$ or states away from the D-branes with very long wavelengths. These two sectors decouple since long wavelength states cannot probe short-distance objects. The fact that these two sectors decouple is in agreement with the gauge theory description in terms of strings between D-branes. The $r \to 0$ states correspond to the short strings between the D-branes, while the long wavelength states correspond to closed strings that do not end on the D-branes, which at low energies are just long wavelength gravitons of type IIB string theory in flat space.

We can study the states near the D-branes with $r \to 0$ more easily by changing to a coordinate

$$u = r\frac{4\pi m_s^2 c^3}{\hbar} \tag{161}$$

which we hold finite as $m_s \to \infty$. This forces $r \to 0$. In this low-energy (near-horizon) limit we find

$$\begin{aligned} ds^2 = \frac{\hbar}{4\pi m_s^2 c^3} \Big[& \frac{u^2}{\sqrt{4\pi g_s N}}(dt^2 + dx^2 + dy^2 dz^2) \\ & + \sqrt{4\pi g_s N}\left(\frac{du^2}{u^2} + d\Omega_5^2\right)\Big], \end{aligned} \tag{162}$$

which is just the metric of a 5-dimensional anti-de Sitter space (AdS_5) and a 5-sphere. Mathematically, since the two spaces are independent (the coefficient of $d\Omega_5^2$ is independent of the other coordinates), the full space is called $\mathrm{AdS}_5 \times S^5$.

Now we can say more precisely what Maldacena's conjecture is: string theory on $\mathrm{AdS}_5 \times S^5$ is equivalent to a 4-dimensional $\mathcal{N} = 4$ supersymmetric Yang–Mills gauge theory. Since the 5-sphere could have a very tiny radius, near the Planck length (141), we are mostly interested in the Kaluza–Klein waves on the sphere[8] that are constant over the sphere. This is because the Kaluza–Klein waves that wiggle on such a small sphere would have masses near the Planck mass (140), and at the energies that are currently available we cannot see any effects of such heavy particles. So Maldacena's conjecture relates string theory in a 5-dimensional anti-de Sitter space to a 4-dimensional conformal theory. Since Maldacena made his conjecture so much circumstantial evidence has been accumulated by comparing the two sides that many people prefer to refer to it as the AdS/CFT correspondence rather than as the Maldacena conjecture. The AdS/CFT correspondence has been extended to many other cases where there is not as much evidence for the equivalence but where there are interesting applications. It seems unlikely that a proof will be found for this correspondence anytime soon, but in the meantime it has been very profitable for theorists to explore the consequences that follow if it is true.

---

[8] Kaluza–Klein waves on a sphere are known as *spherical harmonics*.

If we focus on the anti-de Sitter part of the space–time and change to a new coordinate

$$\rho = \frac{\sqrt{4\pi g_s N}}{u}, \tag{163}$$

then we have a very simple metric for this space:

$$ds^2 = \left(\frac{R}{\rho}\right)^2 \left(dt^2 - dx^2 - dy^2 - dz^2 - d\rho^2\right). \tag{164}$$

We can now see why this theory looks the same on all length scales. Consider what happens when we simultaneously change the length scales of all the coordinates. We can simply multiply all the coordinates by a scale factor so replace the coordinates (t, x, y, z, ρ) by $(\lambda\, t, \lambda\, x, \lambda\, y, \lambda\, z, \lambda\, \rho)$, this means also that $(dt, dx, dy, dz, d\rho)$ are replaced by $(\lambda\, dt, \lambda\, dx, \lambda\, dy, \lambda\, dz, \lambda\, d\rho)$. When we make this replacement in metric (164) we see that the rescaled metric is exactly the same as what we stared with, in other words the metric is invariant under rescalings. Note that this does not happen in the usual metric for a flat space like that given in Eq. (109) which simply gets multiplied by λ^2. It is crucial that both the string theory on anti-de Sitter and the conformal theory had the same symmetries. If they didn't have the same symmetries then we would have known immediately that the Maldacena conjecture was false. Since the basic distinguishing feature of a conformal theory is its conformal symmetry, it is the scale invariance of the anti-de Sitter metric (164) that is the crucial feature to extending the correspondence to different theories.

13.7 Warped Extra Dimensions

In a flat space, the standing wave equation for a wave moving in the x-direction is

$$\left(\frac{\partial^2}{\partial x^2} + k^2\right)\psi(x) = 0, \tag{165}$$

which can be solved for in terms of the trigonometric functions sin and cos since the first derivatives (that is the slope of the function at a given position) of these functions are

$$\frac{\partial}{\partial x}\sin(kx + \phi) = k\cos(kx + \phi), \tag{166}$$

$$\frac{\partial}{\partial x}\cos(kx + \phi) = -k\sin(kx + \phi), \tag{167}$$

which means the second derivatives (the second derivative is the slope of the slope) are

$$\frac{\partial^2}{\partial x^2}\sin(kx+\phi) = -k^2\sin(kx+\phi)\,, \tag{168}$$

$$\frac{\partial^2}{\partial x^2}\cos(kx+\phi) = -k^2\cos(kx+\phi)\,. \tag{169}$$

So the general solution is

$$\psi(x) = a\sin(kx+\phi) + b\cos(kx+\phi)\,. \tag{170}$$

We can then choose a, b, and/or ϕ so that the function ψ satisfied the specified conditions at the boundaries, for example, whether the ends of a string are free to wiggle or not. In fact we only need two of the three parameters a, b, and ϕ to satisfy any boundary condition so we can choose $\phi = 0$, for example, or $b = 0$ as we did in Eq. (88).

The standing wave equation is a special case of the general wave equation

$$\left(\frac{\partial^2}{\partial x^2} - \frac{1}{c^2}\frac{\partial^2}{\partial t^2}\right)F(x,t) = 0, \tag{171}$$

where t is time and c is the speed of the wave. One type of solution is the traveling wave

$$F(x,t) = a\sin(kx-\omega t) + b\cos(kx-\omega t\phi)\,, \tag{172}$$

from which we see that if the time t is increased to $t + dt$ this can be counterbalanced by going to the point $x - \omega dt/k$, so at a slightly later time the wave is equivalent to early wave but now shifted slightly to the right. In fact this solution is a wave that travels to the right at the speed

$$c = \frac{\omega}{k}, \tag{173}$$

which agrees with Eq. (72), (79), and (80). We can see that Eq. (172) is a solution of Eq. (171) since

$$\frac{\partial^2}{\partial x^2}\sin(kx-\omega t) = -k^2\sin(kx-\omega t)\,, \tag{174}$$

$$\frac{\partial^2}{\partial x^2}\cos(kx-\omega t) = -k^2\cos(kx-\omega t)\,, \tag{175}$$

$$\frac{\partial^2}{\partial t^2}\sin(kx-\omega t) = -\omega^2\sin(kx-\omega t)\,, \tag{176}$$

$$\frac{\partial^2}{\partial t^2}\cos(kx-\omega t) = -\omega^2\cos(kx-\omega t) \tag{177}$$

and using Eq. (173) we see that Eq. (171) is indeed satisfied.

Another type of solution for the general wave equation (171) has the form of a function of x multiplying a separate function of t:

$$F(x,t) = A\psi(x)\tau(t). \tag{178}$$

Putting this form in Eq. (171) gives

$$A\left(\tau(t)\frac{\partial^2\psi(x)}{\partial x^2} - \frac{\psi(x)}{c^2}\frac{\partial^2\tau(t)}{\partial t^2}\right) = 0. \tag{179}$$

Dividing by $A\ \psi(x)\ \tau(t)$ we find

$$\frac{1}{\psi(x)}\frac{\partial^2\psi(x)}{\partial x^2} = \frac{1}{c^2\tau(t)}\frac{\partial^2\tau(t)}{\partial t^2}. \tag{180}$$

The left-hand side of this equation depends only on x while the right-hand side depends only on t, the only possibility is that they are both equal to the same constant, which, with some foresight we will call k^2. We now have two equations, and using Eq. (173) we find

$$\left(\frac{\partial^2}{\partial x^2} + k^2\right)\psi(x) = 0, \tag{181}$$

$$\left(\frac{\partial^2}{\partial x^2} + \omega^2\right)\tau(t) = 0. \tag{182}$$

The first equation is just the standing wave Eq. (165) and so we see the solution just has the form given in Eq. (88).

We can do the same type of analysis for standing waves in a 5-dimensional anti-de Sitter space. With the metric given in Eq. (164), where we have called the coordinate of the extra dimension ρ, the standing wave equation for a force carrying field like a photon or Z boson is

$$\left(\frac{\partial^2}{\partial\rho^2} - \frac{1}{\rho}\frac{\partial}{\partial\rho} + k^2\right)\psi(\rho) = 0, \tag{183}$$

which has a solution in terms of the Bessel functions J_1 and Y_1 (see Fig. 13.6)

$$\psi(k,\rho) = \rho\left(aJ_1(k\rho) + bY_1(k\rho)\right). \tag{184}$$

We will take the 2 branes which cut off the extra dimension to be at

$$\rho_{UV} = R = \frac{\hbar c}{10^9\,\text{GeV}} = 2\times10^{-25}\,\text{m}, \tag{185}$$

$$\rho_{IR} = \frac{\hbar c}{246\,\text{GeV}} = 8\times10^{-19}\,\text{m}. \tag{186}$$

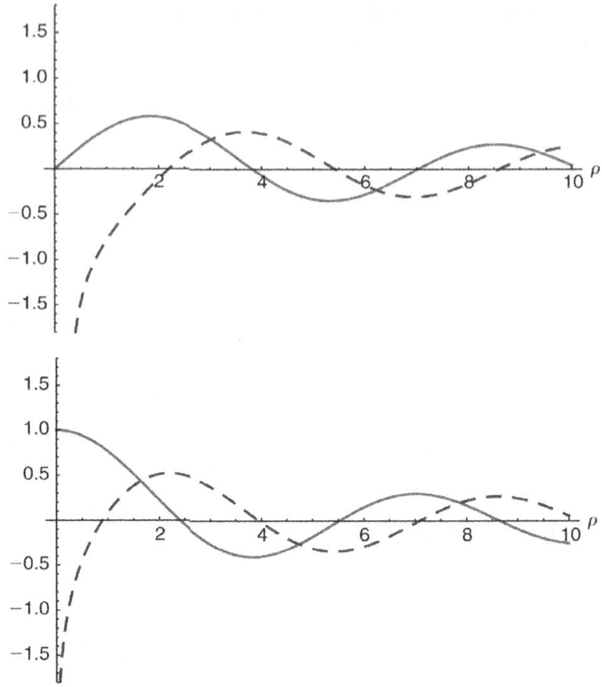

Fig. 13.6 The *solid line* in the *top plot* shows the Bessel function $J_1(\rho)$, while the *short dashes* show the Bessel function $Y_1(\rho)$. The *solid line* in the *bottom plot* shows the Bessel function $J_0(\rho)$, while the *short dashes* show the Bessel function $Y_0(\rho)$

The standing wave is free to wiggle at $\rho = \rho UV$ so we need

$$\left.\frac{\partial \psi(\rho)}{\partial \rho}\right|_{\rho=\rho_{UV}} = k\rho\,(aJ_0(k\rho) + bY_0(k\rho))\Big|_{\rho=\rho_{UV}} = 0 \tag{187}$$

at that brane while if there is a large effective mass on the brane at ρIR then we need the Dirichlet boundary condition

$$\psi(\rho_{IR}) = 0, \tag{188}$$

which we can satisfy by taking

$$b = -a\,\frac{J_1(k\rho_{IR})}{Y_1(k\rho_{IR})}. \tag{189}$$

To satisfy Eq. (187) at $\rho = \rho UV$ (with this value of b) we need

$$\frac{J_0(k\rho_{UV})}{Y_0(k\rho_{UV})} = \frac{J_1(k\rho_{UV})}{Y_1(k\rho_{UV})}, \tag{190}$$

which can be satisfied only for certain values of k. This is how we get a discrete set of standing waves, just as we did for the violin string in Eq. (98). Unfortunately the Bessel functions are more complicated than trigonometric functions, so the solutions for k have to be found numerically using a computer. Using Eqs. (80) and (102) we can find the masses of the particles in this Kaluza–Klein tower. For the values given in Eq. (186) we find

$$\begin{aligned} M_1 &= 90.1\,\mathrm{GeV}/c^2, M_2 = 971\,\mathrm{GeV}/c^2, \\ M_3 &= 2115\,\mathrm{GeV}/c^2, M_4 = 2890\,\mathrm{GeV}/c^2. \end{aligned} \tag{191}$$

The first Kaluza–Klein particle has mass that is just right to be the Z boson (of course I chose the numbers in Eq. (186)) so that this would happen. The next Kaluza–Klein particle is much heavier and would not have been seen in current experiments if it does not couple quite as strongly to electrons and quarks as the Z boson does. However it would certainly be seen at the Large Hadron Collider. This is not a complete model, since we have not accounted for the W boson, but this has been done, it just involves a little more work.

13.8 The Maxwell Equations

Warning: this section is more mathematically intense than the others, and can be skipped if desired.

Maxwell's equations govern the behavior of electric and magnetic fields and their response to electric charges and magnetic dipoles. Electric and magnetic fields turn out to behave very similarly except for one crucial point: while particles can have positive or negative electric charges, no one has ever seen a solitary magnetic charge (also known as a magnetic monopole). Two opposite magnetic charges (that is a hypothetical "north monopole" and a hypothetical "south monopole") could make a dipole like an ordinary bar magnet, but it seems that in nature every magnetic north pole comes with a magnetic south pole because in the absence of magnetic monopoles, all magnetic fields are produced by the movement of electric charges.

In 1784, Charles-Augustin de Coulomb published the results of his experiments on the forces between electric charges. Let's use his results to examine the attractive force between an electron (which has a negative electric charge $-e$) and a proton (which has a positive electric charge $+e$). The form of the force equation is analogous to Newton's universal law of gravitation (119): the force falls off as one over the distance squared and is proportional to the size of each of the charges. Consider an electron a distance r from a proton. To keep track of the direction of the force we can introduce a *vector* $\hat{r}$, which we can think of as an arrow that points from the proton towards the electron. If we put the center of our coordinate system at the position of the proton, then $\hat{r}$ points from the proton position ($x_p = 0$, $y_p = 0$, $z_p = 0$) towards the electron position $\vec{r}_e = (x_e, y_e, z_e)$. More precisely we can write

$$\hat{r} = \left(\frac{x_e}{r}, \frac{y_e}{r}, \frac{z_e}{r}, \right) \tag{192}$$

where we know the distance r is given by $r = \sqrt{x_e^2 + y_e^2 + z_e^2}$. Since the sum of the squares of the components of $\hat{r}$ add up to one:

$$\frac{x_e^2}{r^2} + \frac{y_e^2}{r^2} + \frac{z_e^2}{r^2} = \frac{r^2}{r^2} = 1, \tag{193}$$

the vector $\hat{r}$ is called a unit vector, meaning a vector with length 1. Traditionally we use a "hat" $\hat{}$ to denote a unit vector, and an "arrow" $\vec{}$ to denote a general vector.

Mathematically we can write the (vector) force $\vec{F}$ on the electron as

$$\vec{F} = -\frac{k_C \, e^2}{r^2} \hat{r}, \tag{194}$$

where Coulomb's constant k_C is the analogue of Newton's constant and the overall minus sign reminds us that the force points in the opposite direction to $\hat{r}$, that is from the electron toward the proton. In SI units k_C is defined as

$$k_C = \frac{1}{4\pi \epsilon_0}, \tag{195}$$

where the so-called electric permittivity ϵ_0 has a value

$$\epsilon_0 = 8.85 \times 10^{-12} \, \frac{\mathrm{s}^2 \, \mathrm{C}^2}{\mathrm{m}^3 \, \mathrm{kg}}. \tag{196}$$

Here C stands for the unit of electric charge called a coulomb; in these units the charge of a proton is

$$e = 1.6 \times 10^{-16} \, \mathrm{C}. \tag{197}$$

It's easy to check that with r measured in meters, the units for ϵ_0 give the force in Eq. (194) in units of kilogram meters per second squared (kg m/s^2) which is correct since force equals mass times acceleration. For two electrons (or two protons) the overall sign would be positive and the repulsive force on one of the particles would point along the line between the two particles and away from the other particle.

As early as the 1830s, Michael Faraday realized that it is useful to think of a charged particle as filling the space around it with "lines of force." His conception of lines of force lead to the idea of electric and magnetic fields, and to the modern physics of *field theory*. A single proton produces an electric field given by

$$\vec{E} = \frac{k_C \, e}{r^2} \hat{r}, \tag{198}$$

so that at any point in space the electric field points away from the proton, and the force on a second particle with charge q is given by multiplying the electric field $\vec{E}$ by the charge q:

$$\vec{F} = q\vec{E}. \tag{199}$$

We see that the electric field has units of force per unit charge, which is kilogram meters/per second squared per coulomb (kg m/s^2 C).

The electric field around a proton and an electron is simply the sum of the fields which they would produce alone. Now consider the electric field at the point $\vec{r} = (x, y, z)$ which is a distance r from a proton at (0, 0, 0) and a distance R from an electron at $\vec{r}_e = (x_e, y_e, z_e)$, see Fig. 13.7. The distance R is given by

$$R = |\vec{r} - \vec{r}_e| = \sqrt{(x - x_e)^2 + (y - y_e)^2 + (z - z_e)^2}. \tag{200}$$

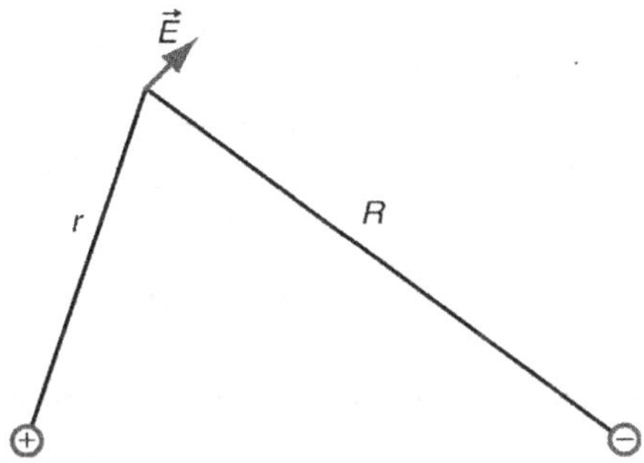

Fig. 13.7 The electric field, $\vec{E}$, at a point near a positive (labeled +) and a negative (labeled −) electric charge. The electric field points in the direction of the arrow; it receives a contribution pointing away from the positive charge and a smaller contribution (since the distance R is longer than the distance r) pointing toward the negative charge. See Eq. (201).

Then the electric field at $\vec{r}$ is

$$\vec{E} = \frac{k_C\, e}{r^2}\hat{r} - \frac{k_C\, e}{R^2}\hat{r}', \tag{201}$$

where we have introduced a new unit vector $\hat{r}'$ that points from the electron to the point (x, y, z). It is a little tricky to work out the direction that the electric field, $\vec{E}$, points in, since it is a combination of $\hat{r}$ and $\hat{r}'$. When we are closer to the proton, $\vec{E}$ mostly points in the direction of $\hat{r}$ and when we are closer to the electron, $\vec{E}$ mostly points in the direction of $\hat{r}'$. The electric field in Eq. (201) is called a dipole field, and it has the same mathematical form as the magnetic field of a bar magnet, so the field points along the field lines shown in Fig. 12.4.

In the presence of both electric and magnetic fields the force on a charge has an extra term compared to Eq. (199). A magnetic field $\vec{B}$ produces a force on a moving

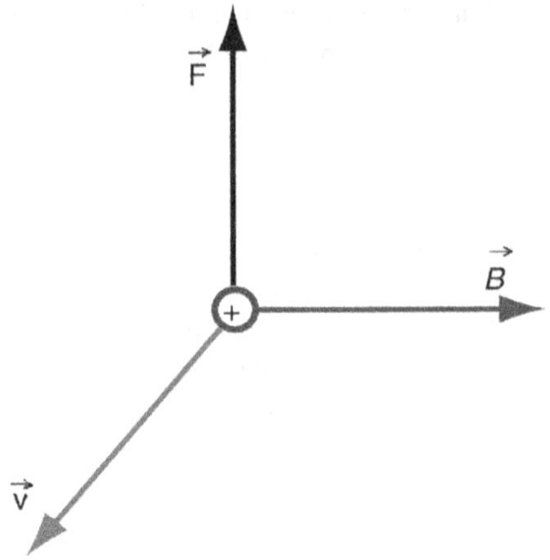

Fig. 13.8 The force, $\vec{F}$, on a proton moving with velocity $\vec{v}$ through a magnetic field $\vec{B}$. Imagine the velocity is coming out of the page, so that all three vectors are orthogonal to each other. See Eq. (204)

charge that has a strength proportional to the strength of the field and the speed[9] of the charged particle. Curiously the force does not point in the direction of the field or the velocity, but in a direction perpendicular to both. Mathematically this direction is given by the *cross-product* of the vectors. For a velocity $\vec{v}$ and magnetic field $\vec{B}$, the x-component of the cross-product is

$$\left[\vec{v} \times \vec{B}\right]_x = v_y B_z - v_z B_y.$$

There are similar expressions for the other components, so we can write the whole vector as follows:

$$\vec{v} \times \vec{B} = \left(v_y B_z - v_z B_y, v_z B_x - v_x B_z, v_x B_y - v_y B_x\right). \tag{203}$$

So for a charge moving through electric and magnetic fields the total force is

$$\vec{F} = q\vec{E} + q\vec{v} \times \vec{B}. \tag{204}$$

This force law was first written down by Maxwell, but has become known as the Lorentz force. Incidentally it tells us that the magnetic field has units of force divided by charge and velocity, which is kilogram per second per coulomb (kg/s C).

The cross-product of two vectors gives a third perpendicular vector, but there is another way to multiply vectors together, called the dot-product, that just gives an ordinary number. For example if we wanted the dot-product of the velocity and the magnetic field we would write

$$\vec{v} \cdot \vec{B} = v_x B_x + v_y B_y + v_z B_z. \tag{205}$$

[9] Speed is the magnitude of the velocity vector.

The dot-product gives us another way to calculate the length of a vector; the length is just the square root of the dot-product of the vector with itself. For example, the length of a magnetic field vector $\vec{B}$ is given by

$$|\vec{B}| = \sqrt{\vec{B} \cdot \vec{B}} = \sqrt{B_x^2 + B_y^2 + B_z^2}. \tag{206}$$

The length of a magnetic field vector directly gives us the strength of the magnetic field.

Writing equations with derivatives in three different directions (x, y, and z) of 3-dimensional vectors can get very complicated as Maxwell found out. Since that time some notational conventions have developed that allow us to write the equations much more compactly than Maxwell could. One of the ingredients we need is a *vector derivative*:

$$\vec{\nabla} = \left(\frac{\partial}{\partial x}, \frac{\partial}{\partial y}, \frac{\partial}{\partial z} \right). \tag{207}$$

The upside-down Δ (delta) is called *del* or sometimes *nabla* (nabla is the Greek word for a type of harp which has a similar shape). Del, $\vec{\nabla}$, acting on a ordinary function of x, y, and z gives a vector that points in the direction of the steepest slope. For a function $f(x, y, z)$, $\vec{\nabla} f$ is called the gradient of f (or *grad* f):

$$\vec{\nabla} f(x,y,z) = \left(\frac{\partial f}{\partial x}, \frac{\partial f}{\partial y}, \frac{\partial f}{\partial z} \right). \tag{208}$$

We can also use $\vec{\nabla}$ to take derivatives of vector functions, like the electric field $\vec{E}$, as well. If we use the dot-product we have

$$\vec{\nabla} \cdot \vec{E} = \frac{\partial E_x}{\partial x} + \frac{\partial E_y}{\partial y} + \frac{\partial E_z}{\partial z}. \tag{209}$$

We call $\vec{\nabla} \cdot \vec{E}$ the divergence of the electric field or *div* for short. We can also use the cross-product to find some complementary information:

$$\vec{\nabla} \times \vec{E} = \left(\frac{\partial E_z}{\partial y} - \frac{\partial E_y}{\partial z}, \frac{\partial E_x}{\partial z} - \frac{\partial E_z}{\partial x}, \frac{\partial E_y}{\partial x} - \frac{\partial E_x}{\partial y} \right). \tag{210}$$

This is called the *curl* of the electric field.

Another way to represent Coulomb's findings is to write

$$\vec{\nabla} \cdot \vec{E} = \frac{\rho}{\varepsilon_0}, \tag{211}$$

where we have related the divergence of the electric field to ρ, the charge density measured in C/m^3. This divergence equation is due to Carl Friedrich Gauss, so it is also known as Gauss' law. For a point charge the solution of Eq. (211) is just Eq. (198). If there are no magnetic monopoles, then there is analogous equation for the magnetic field:

$$\vec{\nabla} \cdot \vec{B} = 0. \tag{212}$$

In 1820 Hans Christian Ørsted found that an electric current[10] in a wire can deflect the magnetic needle of a compass. Within months of the discovery (and just weeks after hearing of it) André-Marie Ampère had carefully analyzed the force between currents in parallel wires and presented his results. Electric current is measured in coulombs per second (C/s) but 1 C/s is now known as one ampere (or 1 amp for short) in honor of Ampère. Equivalent results were also found by Jean-Baptiste Biot and Félix Savart. The experimental findings of all three scientists can be summarized by an equation (Ampère's law) that relates the magnetic field $\vec{B}$ to a steady current density[11] $\vec{J}$ (measured in C/s m^2):

$$\vec{\nabla} \times \vec{B} = \mu_0 \vec{J}, \tag{213}$$

where the so-called magnetic permeability, μ_0, plays a similar role to the electric permittivity ε_0. In SI units its numerical value is

$$\mu_0 = 4\pi\, 10^{-7}\, \frac{\text{m kg}}{\text{C}^2} \approx 1.26 \times 10^{-6}\, \frac{\text{m kg}}{\text{C}^2}. \tag{214}$$

Ampère's finding is the basis behind electromagnets (a current through a wire coil can produce a strong magnetic field) which are the essential feature of electric motors and many other devices.

In 1831 Michael Faraday showed that a magnetic field changing in time produces an electric field. This principle is, of course, used in modern electric generators: moving a wire through a fixed magnetic field produces an electric field which makes a current flow through the wire.

James Clerk Maxwell was able to re-formulate Faraday's result as

$$\vec{\nabla} \times \vec{E} + \frac{\partial \vec{B}}{\partial t} = 0. \tag{215}$$

[10] The electric currents that we use everyday are just made up of the flow of electrons, although no one in Ørsted's time knew about the existence of electrons.

[11] For a wire, the current density is just the current (which is measured in C/s) divided by the area of the wire (measured in m^2). This vector, $\vec{J}$, points in the direction of the positive charge flow. When electrons are flowing the current density points in the direction opposite to the electron flow, since electrons have a negative charge.

Fig. 13.9 Michael Faraday circa 1841

Maxwell was also able to generalize Ampère's law to handle time-dependent fields:

$$\vec{\nabla} \times \vec{B} = \mu_0 \vec{J} + \mu_0 \epsilon_0 \frac{\partial \vec{E}}{\partial t} = 0 \tag{216}$$

The Maxwell equations (211), (212), (215), and (216) unify electricity and magnetism into a single theory of electromagnetism. Maxwell's equations tell us what kind of electric and magnetic fields are produced by charged particles, and the Lorentz force (204) tells us how charged particles move in response to electric and magnetic fields. To understand a particular phenomenon we need to solve all the equations simultaneously, which is usually quite tricky. When quantum effects are included in analyzing such problems (which makes things even more difficult to handle) we have what is known as quantum electrodynamics.

In empty space the Maxwell equations take a simple form:

$$\vec{\nabla} \cdot \vec{E} = 0, \qquad \vec{\nabla} \times \vec{E} = -\frac{\partial \vec{B}}{\partial t}, \tag{217}$$

$$\vec{\nabla} \cdot \vec{B} = 0, \qquad \vec{\nabla} \times \vec{B} = \mu_0 \epsilon_0 \frac{\partial \vec{E}}{\partial t}. \tag{218}$$

Combining the two curl equations Maxwell found that each component of the electric field $\vec{E}$ (that is E_x, E_y, and E_z) and each component of the magnetic field $\vec{B}$ satisfies an equation of the form

$$\vec{\nabla}^2 E_x - \mu_0 \epsilon_0 \frac{\partial^2 E_x}{\partial t^2} = 0, \tag{219}$$

which, as we have already seen, is just the wave equation (171). The Maxwell equations further require that the electric and magnetic fields are perpendicular to the direction the wave is traveling and orthogonal to each other. So for a traveling electromagnetic wave moving in the z-direction, with the electric field pointing in the x-direction we have the solution:

$$E_x = E_0 \cos(kz - \omega t), \tag{220}$$

$$B_y = \sqrt{\mu_0 \epsilon_0}\, E_0 \cos(kz - \omega t), \tag{221}$$

$$E_y = E_z = B_x = B_z = 0, \tag{222}$$

as long as

$$k^2 = \mu_0 \varepsilon_0 \omega^2. \tag{223}$$

From this we can calculate the speed (173) of the electromagnetic waves:

$$c = \frac{\omega}{k} = \frac{1}{\sqrt{\mu_0 \epsilon_0}} = \frac{1}{\sqrt{1.26 \times 10^{-6}\,\frac{\mathrm{m\,kg}}{\mathrm{C}^2} \times 8.85 \times 10^{-12}\,\frac{\mathrm{s}^2\,\mathrm{C}^2}{\mathrm{m}^3\,\mathrm{kg}}}}, \tag{224}$$

$$= 2.99 \times 10^8\,\mathrm{m/s}. \tag{225}$$

This is just the speed of light which had already been measured approximately in Maxwell's time by Hippolyte Fizeau. This was what convinced Maxwell that light is a type of electromagnetic wave. It wasn't until the experiments of Hertz that most of the rest of the scientific community realized that Maxwell was right.

13.9 How Do We Look for Extra Dimensions?

The protons in the Large Hadron Collider beam will have an energy of 7000 GeV, given that a proton has a mass of

$$m_{\mathrm{p}} = 0.938\,\mathrm{GeV}/c^2 \tag{226}$$

we see that the protons have a boost factor of

$$\gamma = 7462. \tag{227}$$

Using Eq. (27) we see that this boost factor corresponds to the protons traveling with velocity that is 0.99999999 of the speed of light.

Steering the protons in a circular orbit is accomplished by using 1232 superconducting dipole magnets, which means that each dipole magnet has to bend the beam by

$$\frac{360^\circ}{1232} = 0.29^\circ. \tag{228}$$

The dipole magnets are built inside of 15 m long tubes, of which 14.34 m is used for bending the beam. Putting all of these 1232 curved segments, each 14.34 m long would make a circle with radius 2812 m and a circumference of 17.6 km. The actual circumference of the Large Hadron Collider is 27 km, since there are also many straight segments included which allow for accelerating regions and detectors.

Applying Lorentz force equation (204) to a relativistic proton one finds that the momentum of a proton (in GeV/c) being bent by a magnetic field B (measured in tesla) in a circle of radius r is

$$p = \gamma mc = 10^{-9}\frac{Br}{c}, \tag{229}$$

so using $r = 2812$ m and $p = 7000$ GeV/c we see that we need an 8.3 T magnetic field. For comparison the Earth's magnetic field is about 0.00005 T while a fridge magnet might be around 0.001 T. Ordinary electromagnets, made from coils of wire (like those found in electric motors), cannot go above 2 T, so to get such an enormous magnetic field the Large Hadron Collider's dipole magnets have to use superconducting wires. To be superconducting the wires need to be cooled to 1.9 K (almost 300 K below room temperature), colder than the microwave background of outer space. As a result the Large Hadron Collider will not only be the highest energy particle colider in the world but also the biggest cryogenics system as well.

About the Authors

Itzhak Bars is a professor of physics at the University of Southern California, Los Angeles. He received his Ph.D. from Yale University in 1971 and after post-doctoral research at the University of California at Berkeley he was appointed to the Faculty of Stanford University in 1973. He returned to Yale University in 1975 as a faculty member in the Physics Department, and after a decade he moved to the University of Southern California in 1984 to build a research group in high-energy physics. He served as the director of the Caltech-USC Center for Theoretical Physics during 1999–2003. His visiting appointments include Harvard University, the Institute for Advanced Study in Princeton, and CERN in Geneva, Switzerland.

Professor Bars is a leading expert in symmetries in physics, which he applies in much of his research on particle physics, field theory, string theory, and mathematical physics in over 200 papers. He is the author of a book on "*Quantum Mechanics*" and co-editor of the books "*Symmetry in Particle Physics*" and "*Strings'95, Future Perspectives in String Theory*." Some of his experimentally successful physics predictions include supersymmetry in large nuclei with even/odd numbers of nucleons, and the weak interaction contribution to the anomalous magnetic moment of the muon, in the context of the quantized standard model, that was confirmed after 30 years. His contributions to the mathematics of supersymmetry are extensively used in several branches of physics and mathematics.

His current interests include string field theory and two-time physics which he originated in 1998. In 2006 he established that all the physics we know today, as embodied in principle in the standard model of particles and forces, is better described by a 2-time field theory in 4-space and 2-time dimensions projected as a shadow on an emergent 3-space and 1-time dimensions. His honors include Fellow of the American Physical Society, the First Award in the Gravity Research Foundation essay contest (shared with Chris Pope), Outstanding Junior Investigator Award by the Department of Energy, and the A. P. Sloan Foundation Fellowship.

John Terning is professor of physics at the University of California, Davis. He received his Ph.D. from the University of Toronto and was a postdoctoral fellow at Yale University. He was also a researcher at Boston University, the University

I. Bars, J. Terning, Founding Editor: F. Nekoogar, *Extra Dimensions in Space and Time*, Multiversal Journeys, DOI 10.1007/978-0-387-77638-5, of California, Berkeley, and Harvard University, and a staff member at the Los Alamos National Laboratory. Professor Terning's research interests include theoretical particle physics, electroweak symmetry breaking, supersymmetry, cosmology, extra dimensions, and the AdS/CFT correspondence.

Index

Q

R

S

T

Made in the USA
Monee, IL
13 September 2026

62976969R00129